NASRULLAH

FORGOTTEN PATRIARCH of the AMERICAN THOROUGHBRED

MELANIE GREENE

THE History PRESS

Published by The History Press
Charleston, SC 29403
www.historypress.net

Cover images: Nasrullah at Claiborne Farm. *Courtesy of Keeneland/Meadors*; Nasrullah. *Courtesy of Keeneland/Meadors*.

First published 2013

Manufactured in the United States

ISBN 978.1.62619.014.6

Library of Congress CIP data applied for.

Notice: The information in this book is true and complete to the best of our knowledge. It is offered without guarantee on the part of the author or The History Press. The author and The History Press disclaim all liability in connection with the use of this book.

For my dad, who fell asleep reading me The Black Stallion *and left me to finish* The Match Race *on my own.*

CONTENTS

FOREWORD

Why write a book about Nasrullah?

Why read one?

The horse, after all, died more than fifty years ago. What's left to say?

Not much, or maybe quite a lot, depending on your point of view.

Everything the statistically minded reader needs to know about Nasrullah probably can be summed up in a handful of pertinent facts: five times the leading sire in North America; first stallion in history to top the sire lists in North America and Britain; sire of numerous champions among nearly one hundred stakes winners; progenitor of a bloodline that persists to this day; and probably the most important imported stallion to stand in North America during the twentieth century.

The statistics are impressive but sterile. Conspicuously absent from the numbers are stories about the horse himself, the flesh on the bone, so to speak.

A case in point:

More than thirty years ago, at the urging of central Kentucky veterinarian and Thoroughbred breeder Dr. E.W. Thomas, I made weekly trips to Paris, Kentucky, to spend afternoons with Colonel Floyd Sager, the resident veterinarian at historic Claiborne Farm. Conducted over the course of several months, those interviews generated a series of articles that ran in the *Blood-Horse* magazine and later were compiled for my first book, *Col. Sager, Practitioner*.

One of the preeminent equine veterinarians of his time, Colonel Sager was at the top of his game during a lengthy tenure at Claiborne Farm. He had in his care dozens of top stallions—Nasrullah among them—and a broodmare band as good as any. He called Nasrullah "one of the most intelligent horses I've been around," but he conceded that the stallion was "a little different."

Calling Nasrullah "different" was being kind. The horse was a rank troublemaker when he raced in England and during his stud career at Claiborne.

"He was smart," Colonel Sager recalled, "but he seemed to use all his intelligence to make trouble for someone, and that someone usually was me. He would kick or strike, anything to keep you awake when you were around him. And if he didn't want to do something, then he didn't do it.

"It almost seemed like Nasrullah would hold a grudge against you. I could go to his stall where he was just standing and eating hay, and as long as I didn't say anything, he was all right, even if there were a half dozen other people around. But as soon as I spoke, old Nasrullah would rear and whirl around and lash out at the old screen on his door and just generally raise hell."

On another ill-fated occasion, Colonel Sager asked Nasrullah's stud groom to put a twitch on the horse. Things didn't go well.

"So we back Nasrullah up in the corner of one of those stalls," Colonel Sager recalled, "and Lawrence got the twitch on him, but the instant that twitch tightened, Nasrullah sat down in the corner and lunged out. I never had gotten close to the horse, and Lawrence just turned him loose. Nasrullah went across that stall, and he left marks from his front feet for about six or eight feet along the ceiling, then he crashed into the wall on the opposite side."

Colonel Sager's recollections of Nasrullah may not be integral to an understanding of the horse's lasting influence on Thoroughbred racing. The same can be said of author Melanie Greene's stories of His Highness the Aga Khan, who bred and raced Nasrullah during the height of World War II, or of Bull Hancock, whose prescient judgment led him to bring the horse to this country. The mere fact that Nasrullah's name still can be found in the pedigrees of many extraordinary horses, albeit a few generations back, is testament to his lasting impact on the sport.

The stories are important nonetheless, important to an understanding of another time, when racing was more about the horses and less about the business.

Which brings us back to the original question.

Why take the time to delve into the history of a horse that's long dead and buried?

At least two reasons come to mind.

Historians like to discover connections and weave a complete tapestry from dusty file boxes crammed full of isolated facts. For the best historians, the ones who make their subjects come alive, the whole of a story really is greater than the sum of the parts. *Good* history is *accessible* history, and that's where books like *Nasrullah* play a significant role.

More important, though, is the need to preserve a collective memory that is fast becoming lost. While researching my books about Dancer's Image and Noor, I was dismayed to realize how much of racing's rich history is on the verge of extinction. The digital age is both a blessing and a curse. Archives can live forever as digital files stored on a server somewhere or in the "cloud," whatever that really is, but documents and photographs that don't make the cut for immediate digitization deteriorate over time. Sometimes, tragically, invaluable material is simply thrown away with the trash.

People pass on, their memories die with them and old photographs fade away, often without anyone noticing they're gone until it's too late.

The *Thoroughbred Times*, for years one of the two major weekly trade journals covering Thoroughbred racing, went into bankruptcy in September 2012. The assets of the publication, including an extensive library and many thousands of photographs and news clipping files, were auctioned online. The Keeneland Library, one of the world's leading repositories for racing's history, was the successful bidder for the books, photographs and files, and the magazine's archives are safe.

The Keeneland Library also has undertaken the monumental task of digitizing its irreplaceable collection of *Daily Racing Form* and *Morning Telegraph* newspapers. A handful of other museums have significant collections, and a few individuals—Kip Hannan at Hollywood Park and veteran racing photographer Bill Mochon come immediately to mind—are working to preserve racing's film and photographic history. The neverending task is too big, though, and too important, for just a handful of dedicated people.

So, why read a book about Nasrullah?

Because Thoroughbred racing is a product of its past, both figuratively and literally, and because the past matters. When the *Blood-Horse* polled experts to identify the most important events in Thoroughbred racing that occurred during the twentieth century, the importation of Nasrullah to the United States ranked twenty-third among the top one hundred. That recognition, standing alone, is reason enough to thank Melanie Greene for committing Nasrullah's story to paper.

The fact that Greene's *Nasrullah* is an entertaining read is a pleasant bonus, like a winning pari-mutuel ticket discovered, uncashed, in a jacket pocket.

MILTON C. TOBY
Author of *Dancer's Image: The Forgotten Story of the 1968 Kentucky Derby* and
Noor: A Champion Thoroughbred's Unlikely Journey from California to Kentucky
Georgetown, Kentucky
February 2013

PREFACE

Secretariat, Ruffian, Mr. Prospector, Nashua, Bold Ruler, Noor...these equine legends—as well as a great many more—immortalized by their heroic exploits on the track, descend from the mighty stallion Nasrullah. Without his blood coursing through the veins of thousands of Thoroughbreds born each spring, the American horse racing industry would be a very different one indeed. Despite his profound influence, however, the story of Nasrullah remains relatively unexplored in historical accounts of racing and breeding, beyond brief mentions by pedigree analysts, and he is largely unknown to the average racing enthusiast.

Perhaps one reason that the story of Nasrullah has not been recounted is the fact that media attention tends to focus on, and draw inspiration from, sports heroes and underdogs; Nasrullah was neither, at least not in the way one would imagine. In horse racing, the flashes of brilliance that horses show on the track, and what they overcome to get to the winner's circle, is what is remembered and what is preserved.

People outside of the racing industry, and even some of those directly involved in it, do not know a lot about the centuries-old traditions of breeding, the theory of "breed to the best and hope for the best," the intricacies and uncertainties of equine genetics, the patience it takes to wait decades for the "right horse" and the level of dedicated horse care that is responsible for bringing such athletes to life at the track. There is certainly just as much mystery, magic and delight, however, to be found in uncovering the history of these horses and the lives of the people who brought them to us.

Nasrullah was certainly no underdog, nor is he remembered as a hero on the racetrack. Yet he was regally bred, and the best people in the horse business were associated with, and largely responsible for, this stallion's success, which was brought to the racing world via the breeding shed.

I became enthralled with the untold story of Nasrullah after years of examining pedigree charts, bloodlines, stud registries and books that hinted at his role in shaping the modern American Thoroughbred. As a child, I would spend hours reading about the great horses that carried his blood without taking notice of their ancestry. As an adult, I continued to see his name crop up as I probed the pedigrees of horses I rode and watched race, and I wondered, "Who was this horse, and why hasn't anyone told his story?"

The more I read, the more fascinated I became, and eventually it turned into an obsession, distracting me from the writing of my doctoral dissertation and at times threatening to delay the completion of my degree, unbeknownst to my academic supervisor. It became so important to me to share with readers the mystery of this horse, the legacy of which lives on today and will continue to live on for years to come, in the bloodlines of horses we see at racetracks, at horse shows and grazing in pastures around the country and the world.

Through my research, I have discovered that Nasrullah and I share an interesting connection. In the early twentieth century, Nasrullah's breeder, Sir Sultan Mahomed Shah, His Highness Aga Khan III, was in India serving as president of the All India Muslim League, a political party dedicated to creating an independent Muslim nation in northwest India. At the same time, my great-great-great-grandfather Sir Henry John Stedman Cotton was serving there in the Indian Civil Service. In 1904, he served as president of the Indian National Congress, one of the few non-Indians to do so. Cotton was sympathetic to Indian nationalism and supported Indian Home Rule. He advocated the cause in his 1885 book, *New India, or India in Transition*, for which he was criticized. It is certainly probable that the paths of these two leaders might have crossed at some point in their shared commitment to India's independence.

As the champions of the twentieth century recede further and further from the memories of people currently involved in horse racing, and as the sport itself continues to face change and uncertainty, it becomes more and more important that the stories of yesteryear's champions are documented for future generations of equine and sports enthusiasts to learn about and to share.

It has been suggested that Nasrullah had the greatest influence of any of the imported North American stallions of the twentieth century. Long after his own hoof beats have faded from memory, the sound of his descendants' hooves can be heard on tracks all over the world. This is his story. I hope you enjoy it as much as I enjoyed the journey of researching and writing it.

ACKNOWLEDGEMENTS

I would like to thank the following individuals and organizations for their contributions to this work:

Cathy Schenck at Keeneland Library, in Lexington, Kentucky, who provided invaluable resources and photographs of Nasrullah;

Dr. Peter Physick-Sheard at the Ontario Veterinary College, University of Guelph, who patiently explained equine cardiology in a nutshell;

Alan Grundy at the National Horseracing Museum, Newmarket, Suffolk, United Kingdom, for providing photographs;

Gerald Segasby, who provided the photographs of Nearco, Nearula, Never Say Die, Mill Reef and Zucchero;

Linda Rainey at the Canadian Horse Racing Hall of Fame, in Toronto, Ontario, for her assistance in locating and providing photographs of Nasrullah's Canadian descendants;

Arianna Galuzzo at the National Museum of Racing and Hall of Fame, Saratoga Springs, New York, who provided the photograph of Bold Ruler;

The staff at Memorial University of Newfoundland's Digital Media Centre, who assisted with scanning and photo editing;

Coffee Matters café for providing a pleasant setting for writing and delicious sandwiches;

Kimberly Gatto for her encouragement and support of this work;

Milton Toby for enthusiastically agreeing to write the foreword;

The staff at The History Press, especially my commissioning editor, Banks Smither, and publicist Katie Parry;

Erika Steeves, who reviewed and edited the manuscript and offered feedback and suggestions;

Thomas Ozon, for his help with the index;

My four-legged friends Holly and Arrow for their patience and inspiration.

Chapter 1

INTO ROYALTY A KING IS BORN

The year 1940 was one of triumph and turmoil. For the United Kingdom, it started off with tragedy—a foreshadowing of sinister events to come. On January 9, the British submarine HMS *Starfish* was sunk by a German minesweeper at Heligoland Bight, and its crew was captured. It would be the first of many attacks at sea that year.

At the start of World War II, Ireland was a dominion of the British Empire and a member of the Commonwealth. The Anglo-Irish Treaty of 1921 declared Ireland's independence; as a result of the Statute of Westminster, Britain's involvement in the war did not automatically include its dominions. Ireland was thus able to maintain its policy of neutrality throughout the duration of the war. But that did not ensure the safety and peace of its people.

On February 7, the passenger ship *Munster* became the first Irish ship to be sunk in World War II, after it was mined in Liverpool Bay. Over two hundred passengers and the crew of fifty, including commander Captain J. Paisley, escaped in lifeboats and were rescued a few hours later by the collier *Ringwall*.

Overseas, happier times were being had in the United States. The very first McDonald's restaurant opened on May 15 in San Bernardino, California. It wouldn't be until 1979, however, that the first Happy Meal appeared. Sport was forever changed with the first television broadcasts. The first televised hockey game was broadcast on February 25, followed by the first televised basketball game three days later. On April 13, the

New York Rangers won the Stanley Cup for the third time, beating out the Toronto Maple Leafs. It would be another fifty-four years before they tasted victory again.

A pivotal moment in American racing came on April 12, opening day at Jamaica Racetrack, which featured the use of parimutuel betting equipment. Other New York tracks quickly followed suit. Tote betting, as it is known in the United Kingdom, had been introduced there eleven years prior by Winston Churchill, who established the Racehorse Betting Control Board in 1928. The first major flat-race meetings with tote betting were at Newmarket, on the July Course, in 1929.

Amidst political and societal turmoil, a brown Thoroughbred colt was born on March 2, 1940, at Sheshoon Stud in County Kildare, Ireland. He was named Nasrullah, an Arabic name meaning "victory of God." Sheshoon Stud is the oldest continuously operating stud in England or Ireland. It was also the first Irish farm purchased by the third Aga Khan, spiritual leader of the Ismaili sect of Shia Muslims.

The influence of the Aga Khan family on Thoroughbred breeding over the past century or so has been immense. Tony Morris, of the *European Bloodstock News*, was quoted as having said he "would be prepared to bet a substantial sum that there is no notable thoroughbred in the world without some example of Aga Khan breeding in its background."[1]

Sultan Mahomed Shah was born in Karachi, Pakistan, in 1877. He inherited the title of Aga Khan III in 1886 at just eight years of age, after his father's sudden death from pneumonia. Not long afterward, he received the title of His Highness from Queen Victoria.

The Aga Khan's grandfather, Aga Hasan Ali Shah, Imam of the Shia Imami Ismaili Muslims, was given the title of Aga Khan I by the emperor of Persia, Fat'h Ali Shah Qajar. The name Aga Khan derives from the Turkish military title *agha* combined with the Turkic, Mongolian or Persian/Pashto title *khan*, translated as "commanding chief."

Aga Khan I had been governor general of the Persian province of Kerman before a dispute with the shah led him to seek the protection of the British in India. His eldest son, Aqa Ali Shah, succeeded his father in 1882, but his time as Aga Khan II was short-lived, lasting only four years and ending with his untimely demise.

The dynasty of the Aga Khan shares a long and cherished history and affinity with the horse. In his memoir, His Highness Aga Khan III recalls that his earliest childhood memory is of sitting astride his pony alongside his grandfather, the first Aga Khan, seated on a gray Arab mare, watching his horses in training.[2] He stated that his grandfather was not the first member of his family to take an interest in horses, "a passion that historians have traced back to sixth-century Arabia."[3] Philip Jodidio, writing in *A Racing and Breeding Tradition: The Horses of the Aga Khan*, notes:

> *The history of the racing and breeding activities of the Aga Khan family can be traced back to nineteenth-century India, although the foundation of the current operation is more related to the purchases of His Highness Aga Khan III (1877–1957) in the 1920s.*[4]

Upon the death of his father, Aga Khan III inherited his family's racing stables; his horses raced under his name and met with great success on the turf of western India. In the 1933 book *Prince Aga Khan: An Authentic Life Story*, Sirdar Ikbal Ali Shah described the Aga Khan's love of horses:

> *His Highness has inherited from his forefathers a great and genuine love of horseflesh. He can appraise the possibilities of an animal at a glance, and sometimes prefers to rely more upon his own judgment in these matters than upon that of others. He is not an owner who takes but a cursory interest in his actual stables. He displays the utmost respect for every animal that is called upon to carry his green and chocolate hoops. More frequently than not, it is he who says how a difficult candidate should be ridden, in order that the best might be got out of him.*[5]

It is thus unfortunate that the Aga Khan was unable to take an active participation in the training and racing career of Nasrullah, as his expertise most certainly would have come in handy with the difficult colt; alas, as a result of the war, the Aga Khan was sent into exile in Switzerland during the horse's formative years.

Aga Khan III had been encouraged to give breeding and racing in England a try by a Mrs. Asquith, daughter of the former prime minister Herbert Henry (H.H.) Asquith and a sister-in-law of George Lambton, trainer for Lord Derby. The two engaged in conversation at a dinner party hosted by Mrs. Edward Montagu in London. On her advice, the Aga Khan wrote to Lambton, who introduced him to trainer Richard (Dick) Dawson, and agreed to purchase some yearlings on his behalf.

Lambton was unable to fulfill His Highness's request to train for him, as he already had a full yard. Instead, he became responsible for the purchases of the first of the Aga Khan's foundation mares, including Nasrullah's

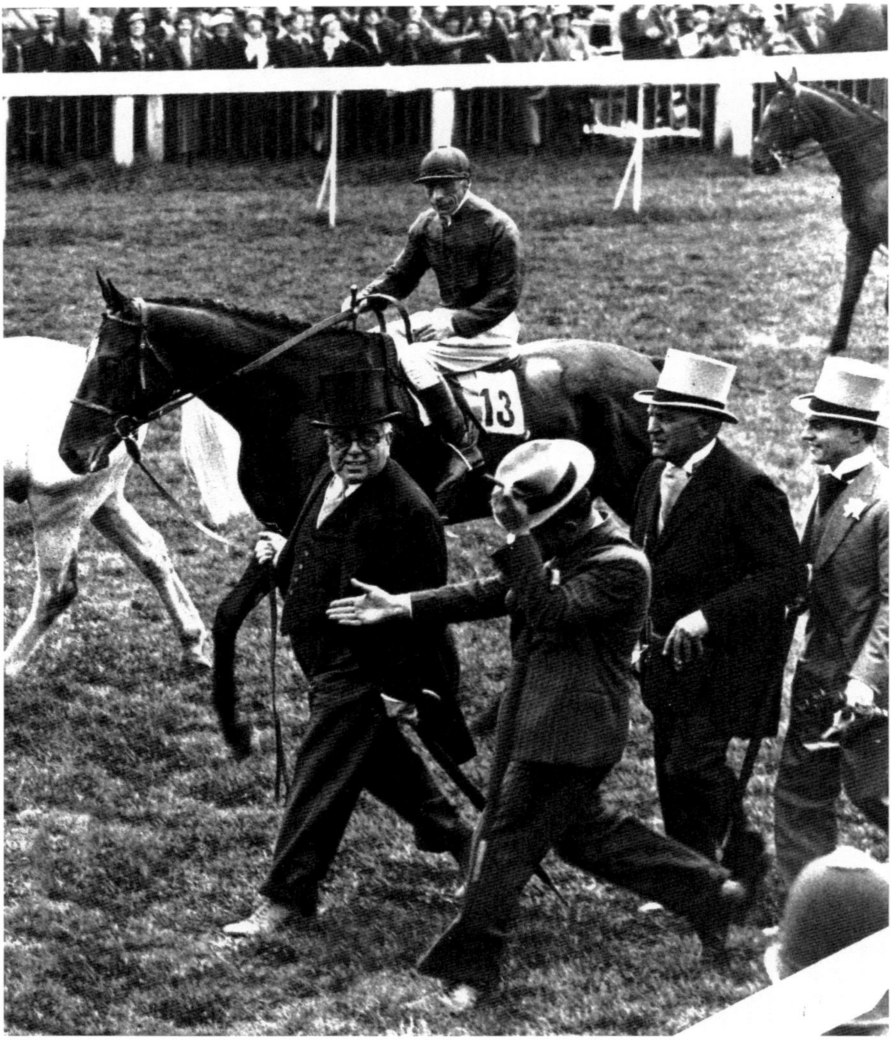

The Aga Khan III, Nasrullah's breeder and first owner, leads in Bahram, winner of the 1935 Epsom Derby. *Courtesy of the National Horseracing Museum, Newmarket, United Kingdom.*

grand-dam Mumtaz Mahal. As a result of the filly's wins as a two-year-old, as well as the racing success of Diophon, Salmon-Trout, Teresina and Paola, the Aga Khan was second on the leading owner's list in 1923. The following year, he led the list for the first of thirteen times. He was also leading breeder for eight seasons. Among the first of his sixteen Classic wins were the 1924

2,000 Guineas with Diophon and that year's St. Leger with Salmon Trout. In 1926, he got the winner of the Irish Derby with Zionist.

In 1930, the Aga Khan got his first Epsom Derby win with Blenheim, Nasrullah's maternal grandsire. This was followed up with 1932 Classic wins by Udaipur (Oaks) and Firdaussi (St. Leger). The Aga Khan returned to the derby winner's circle two years in a row, leading in Bahram (1935) and Mahmoud (1936).

In 1939, His Highness made an offer to the Italian government to sell all of his horses, and while he had considerable support from people of high standing, Mussolini forced him to abandon the deal. Prior to this, the British National Stud had rejected the same offer to buy his entire stable, for reasons unknown to the Aga Khan.[6] One is left to wonder what the fate of the then unborn Nasrullah would have been if either of these deals had gone through.

While many moments from the year 1940 remain forever frozen in time, Nasrullah's birth went relatively unnoticed. Today, his life has been largely forgotten, and the extraordinary legacy he left behind remains unacknowledged and underappreciated. Unbeknownst to those few who witnessed his emergence into the world, horse racing would forever be changed. He arrived in the relative calm before the storm of World War II hit Britain's shores. A little over two weeks after his birth, on March 18, Adolf Hitler and Benito Mussolini met at Brenner Pass in the Swiss Alps and agreed to form an alliance against France and the United Kingdom. The world would never be the same.

Chapter 2

ECHOES OF ANCESTORS

To forget one's ancestors is to be a brook without a source, a tree without root.
—Chinese proverb

Nasrullah descends from the Sir Hercules line of Eclipse, arguably the most dominant line from which leading American Thoroughbred sires trace. Nasrullah also traces 5x5x6x6 to St. Simon, hailed as the greatest horse of his crop. Born in 1881, St. Simon was sired by Galopin, winner of the Epsom Derby and three-time leading sire in Great Britain and Ireland. Winner of the Ascot Stakes, Ascot Gold Cup, Epsom Gold Cup and Goodwood Cup, St. Simon retired undefeated and went on to be champion sire nine times, as well as leading broodmare sire six times in Great Britain and Ireland. His progeny won a total of seventeen English classics.

Maternal Line

Nasrullah's royal lineage traces back to memorable matriarchs. His dam, the French-bred Mumtaz Begum, while not an overly successful racehorse, winning only two races, was out of Mumtaz Mahal, recognized as one of the most important broodmares of the twentieth century.

Mumtaz Mahal

Known by track-goers as the "flying filly," and affectionately as "Mumty," Mumtaz Mahal inherited both the unique spotted coloring and speed of her sire, The Tetrarch, but not the middle-distance abilities of her dam, Lady Josephine. Winner of the Coventry and Acorn Stakes races, Lady Josephine was a daughter of sprinters Sundridge, champion sire in Great Britain in 1911, and the mare Americus Girl, whose sire, Americus,[7] was a stakes winner in the United States.

Bred by Lady Sykes at Sledmere Stud in Driffield, in the East Riding of Yorkshire, Mumtaz Mahal was bought at the Doncaster yearling sale in 1922 for 9,100 guineas—the second-highest price ever paid for a yearling at that time—by the Honorable George Lambton on behalf of His Highness the Aga Khan III. Formerly an amateur steeplechase jockey, Lambton was champion trainer in the 1906, 1911 and 1912 seasons and trained many horses for British flat racing's noble and elite, including the winners of ten classics for Lord Derby.

Named for the Mughal empress, for whom the Taj Mahal was built, Mumtaz Mahal won seven of ten races, set a track record in the Spring Two-Year-Old Stakes at Newmarket in 1923 and was voted champion two-year-old filly in England. As a three-year-old, Mumtaz Mahal won the King George Stakes and Nunthorpe Stakes, both in the company of colts.

As a broodmare at Sheshoon Stud, she produced nine foals, six of them winners, including Mirza II, which Frank Butters claimed to be the fastest horse he ever trained; the filly Mah Mahal, which later produced Epsom Derby winner Mahmoud; Mah Iran, a top two-year-old filly that produced Migoli, winner of the Prix de l'Arc de Triomphe and sire of Belmont Stakes–winner Gallant Man; and Star of Iran, dam of the Aga Khan's great filly Petit Etoile.

In 1931, Mumtaz Mahal was bred to the Aga Khan's Blenheim II. The result of the mating was Mumtaz Begum, dam of Nasrullah. Mumtaz Begum was barren her first season at stud, and then she produced Sun Princess, a filly by Solario that became the dam of Royal Charger, a very successful sire. She was next bred twice to Dastur, producing two fillies. Nasrullah was her fourth foal and the first colt. In all, she had ten foals, eight of them winners, including Nasrullah's full siblings, Rivaz, Nizami—not to be confused with the other Nizami (Firdaussi-Mumtaz Mahal)—and Malindi, all of them stakes winners.

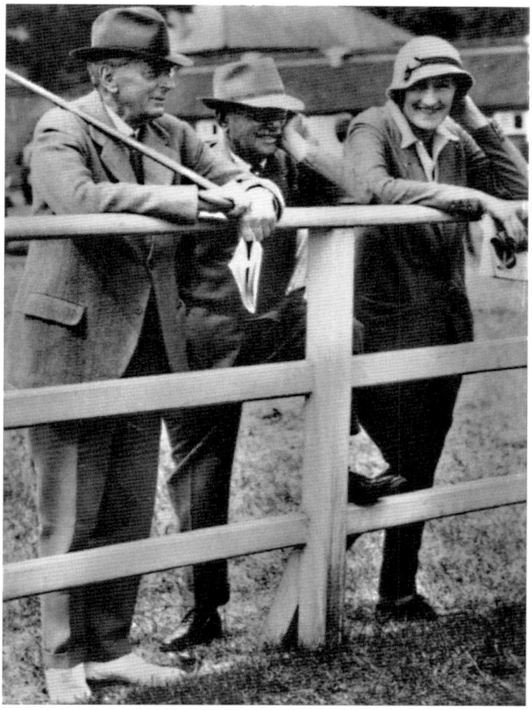

Left: The Honorable George Lambton (far left), who purchased a number of horses on behalf of the Aga Khan III. *Courtesy of the author.*

Below: Mumtaz Mahal, the "Flying Filly," Nasrullah's grand-dam. *Courtesy of the National Horseracing Museum, Newmarket, United Kingdom.*

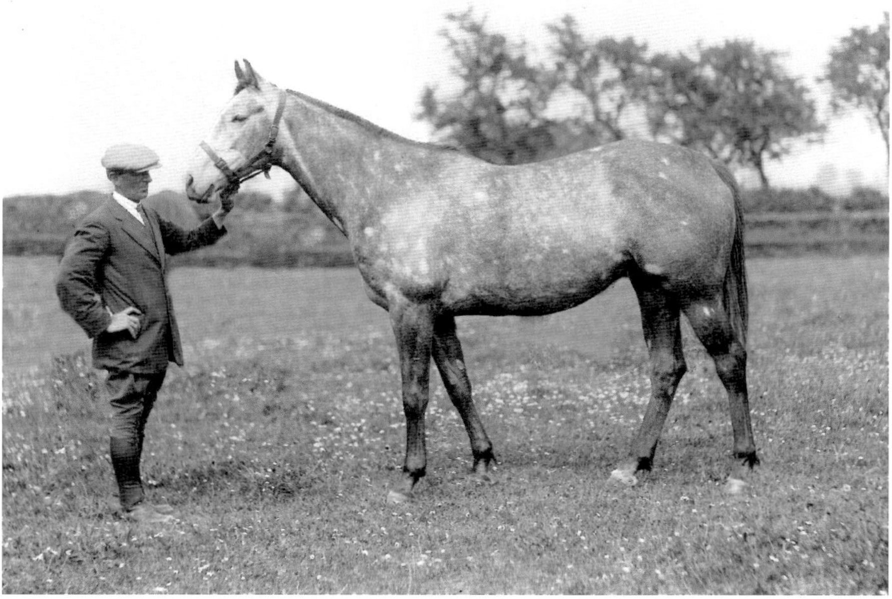

The Tetrarch

Mumtaz Mahal's sire, The Tetrarch, was bred by Edward Kennedy, sired by the French stallion Roi Herode and out of the English mare Vahren, whose sire, Bona Vista, won the 2,000 Guineas. He was foaled at Straffan Station Stud near Ardclough in County Kildare. The Tetrarch was a gangly and awkward foal with very distinctive coloring. He was described as being "a kind of elephant gray with white and lime patches…as if someone had splashed him with a brush dipped in a bucket of whitewash."[8] He was dismissed by buyers and considered to have little potential. Finally purchased by Major Dermott McCalmont, he was sent to Atty Perse for training. Initially laughed at by race-goers for his appearance when he arrived at Newmarket in 1913, the colt quickly put his doubters to shame when he "exploded away from his 22 rivals to win his debut, a maiden two-year-old plate, by four lengths in a canter."[9]

Ridden by leading jockey Steve Donoghue, The Tetrarch was unbeaten at two, winning seven races and earning the nicknames the "Spotted Wonder" and the "Rocking Horse." He was crowned champion juvenile colt in

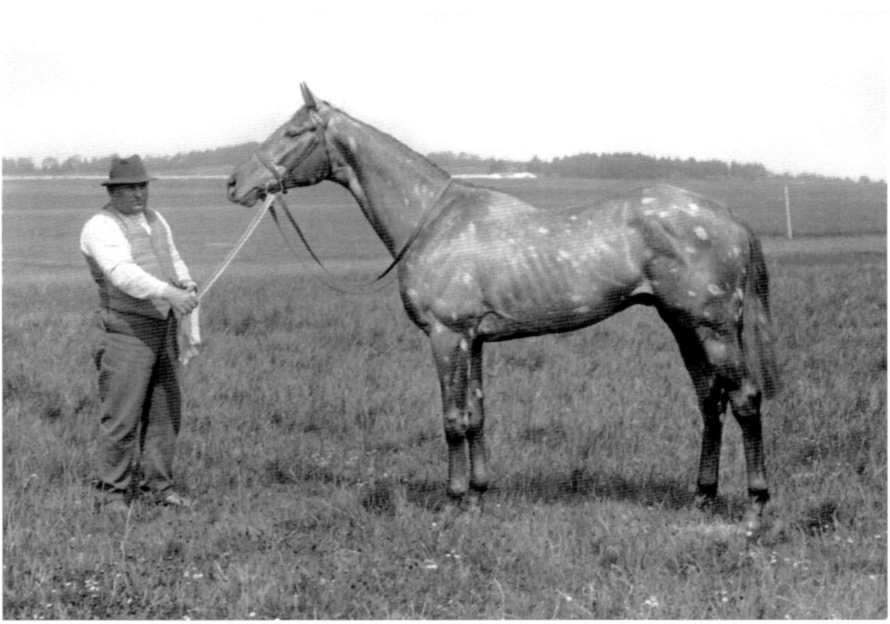

The Tetrarch, the "Spotted Wonder." *Courtesy of the National Horseracing Museum, Newmarket, United Kingdom.*

England, and the National Horseracing Museum in the United Kingdom later voted him Britain's two-year-old of the twentieth century. As a result of repeated injuries, however, The Tetrarch was retired prematurely to stud. As he never raced beyond a distance of six furlongs, it is uncertain what he might have achieved when he was older or been capable of at longer distances. He had plenty of stayers in his pedigree, and his progeny won at varying distances ranging from sprints to classic distances.

Sent to Thomastown Stud, The Tetrarch had difficulty as a stallion and was plagued by fertility problems for most of his stud career. Although he sired only 130 foals, he was leading sire in Great Britain and Ireland in 1919 due in part to the racing performances of Tetratema and Snow Maiden. Three of his sons won the St. Leger Stakes: Caligula, Polemarch and Salmon-Trout. He was retired to McCalmont's Ballylinch Stud and died on August 8, 1935, at age twenty-four.

Blenheim II

Blenheim II, bred by Lord Carnavon at Highclere Stud, was purchased by the Aga Khan as a yearling for 4,100 guineas. Blenheim II was by Blandford, which was bred by the National Stud, led the sire list three times and produced eleven English classic winners. He remains one of only four horses to sire four winners of the derby; the others are Waxy, Sir Peter Teazle and Cyllene. Blandford's other three derby winners were Trigo (1929), Windsor Lad (1934) and Bahram (1935). Blandford was a son of St. Leger winner Swynford, which George Lambton claimed was the best horse he had ever trained. Swynford sired six classic winners, including derby winner Sansovino, and was leading sire in 1923.

Winner of four races (all at a distance of five furlongs) out of seven starts as a two-year-old, Blenheim II also claimed the 1930 Epsom Derby, the first of three derby victories for the Aga Khan.

After his racing career, Blenheim II went to stud in France at the Aga Khan's Haras Marly La Ville in 1931. His French-sired crops included derby winner Mahmoud, Italian champion Donatello II, French 2,000 Guineas winner Drap d'Or and French 1,000 Guineas winner Blue Bear, as well as Mirza II, a full sister to Mumtaz Begum and thus Nasrullah's "aunt." In 1940, the year Nasrullah was born, Mirza II was stolen by the German army and brought to Heeresgestüt Altefeld, which was used for racing and breeding by the German armed forces at the time. After the war, she was sent to England.

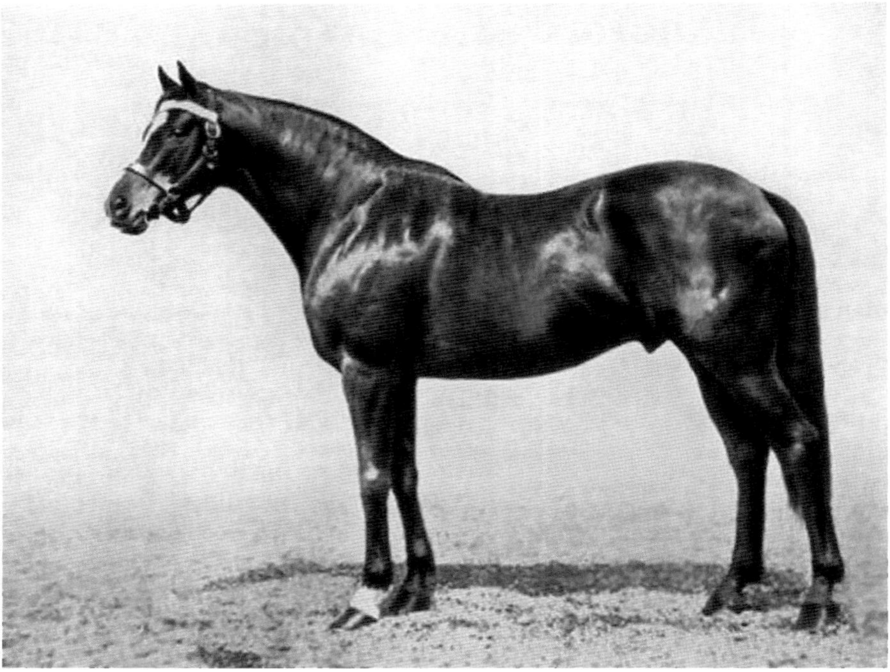

Blandford, three-time leading sire in England and sire of eleven English classic winners. *Courtesy of Wikipedia Commons.*

Blenheim was later sold for $225,000[10] to an American syndicate consisting of Warren Wright (Calumet Farm), Arthur B. Hancock Sr. (Claiborne Farm), William DuPont, Mrs. Thomas H. Somerville, J.H. Whitney (Greentree Stud), John D. Hertz (Stoner Creek Stud) and Robert A. Fairbairn (Fairholme Farm). He was imported after the 1936 breeding season to stand at Claiborne Farm.

Blenheim had great success as a stallion in America, leading the sire list there in 1941. He sired American Triple Crown–winner Whirlaway; champion handicap mare Mark-Kell; her full sister, Kentucky Oaks–winner Nellie L. (both out of Nellie Flag); and Kentucky Derby–winner Jet Pilot, among others. He also made his mark as a broodmare sire. Influential daughters included Easy Lass (Broodmare of the Year and dam of champions Coaltown and Wistful), Jane Gail (dam of Kentucky Derby–winner Hill Gail), Miss Rushin (dam of Kentucky Derby–winner Pensive) and Sweep In (dam of Kentucky Derby– and Preakness-winner Kauai King).

Paternal Line

Nearco

Nasrullah descends from a long line of kings. His sire was the Italian stallion Nearco, undefeated in fourteen starts. Sire of other noteworthy stallions Nearctic and Royal Charger, Nearco is responsible for both the Northern Dancer and Bold Ruler lines, arguably the most influential in the modern Thoroughbred.

With the exception of three stallions—Exclusive Native, Halo and Saint Ballado—as of 2011, all U.S. general sire list leaders from the Sir Hercules line of Eclipse possess the blood of Nearco, either through Nasrullah (Bold Ruler line) or Nearctic (Northern Dancer line). His impact on the breed is thus irrefutable, and the genius of Federico Tesio is largely responsible.

The tale of how Nearco came to be is an interesting story in itself. As Nearctic biographer, Muriel Lennox, recounts:

> *Nearco is the proof of human folly, a reminder that the only certainty in the breeding of great Thoroughbred champions, is uncertainty. If Tesio, recognized as the breeder of Nearco, had had his way, there would not have been a Nearco.*[11]

Federico Tesio bred some of the world's most famous racehorses at his Dormello Stud on the banks of Lake Maggiore in Northern Italy. His other great horse was Ribot. Tesio was very passionate about horses and became even more fascinated with the fundamentals of equine genetics and breeding when he had a chance encounter with the infamous Gregor Mendel on a train traveling from Pisa to Rome. Tesio read of the Austrian botanist's experiments in crossing pure breeds and developing hybrids, and his highly regarded *Breeding the Racehorse* is the result of his application of Mendel's theories to the Thoroughbred.[12]

Tesio got the regally bred Catnip for a steal, paying 75 guineas for her at the December 1915 Newmarket sales, a purchase that certainly paid off for the Italian breeder. The Irish-bred mare was a daughter of Epsom Derby–winner Spearmint[13] and the American-bred mare Sibola, winner of the 1,000 Guineas and Oaks runner-up. Catnip's paternal grandsire was the New Zealand–bred Hall of Fame sire Carbine, winner of the Melbourne Cup and other stakes in Australia.

When Tesio purchased Catnip, she was in foal to Cock-a-Hoop. Sadly, the foal died, and the following year the mare was barren to the stallion Lemberg. Her next foal was Nera di Bici, considered to be the best filly to come from Dormello. Catnip was next mated with Havresac III (by Rabelais, winner of the Goodwood Cup and three-time leading sire in France), which led the Italian sire list eleven times. Tesio had high hopes for their foal, Nogara, and she certainly lived up to them.

Nogara was a sprinter-miler and won two Italian classics, including the Premio Regina Elena, the Italian version of the 1,000 Guineas. As a broodmare, she was a blue hen and black type producer. She produced nine foals, including three classic winners in Nearco, Niccolo Dell'Arca (Italian Derby) and Naucide (Italian Oaks).

After much consideration, Tesio had decided to breed Nogara to Fairway, winner of the St. Leger and champion three-year-old colt in England in 1928. However, by the time he finally made up his mind, the stallion's studbook was full. Reluctantly, Tesio sent Nogara to France to be bred to Pharos, Fairway's full brother. Lucky that he did, for if he hadn't, there would not have been a Nearco and therefore no Nasrullah.

Nearco was born on January 24, 1935, at the Dormello Stud. As a foal, Nearco was small and stocky, and he did not initially impress Tesio. Nearco proved to be a willful sort, however, bullying other foals in the paddock and exhibiting aggressive behavior that was encouraging to Tesio. As a weanling, Nearco was shipped south to Olgiata, the estate of Mario Incisa della Rochetta—a successful migration experiment that today is replicated by many breeders in North America, who often send their young horses south to mature and then begin training.

Nearco was a strong and gutsy yet good-tempered horse. He won seven races as a two-year-old, including the Criterium Nazionale, Gran Criterium, Premio Teveres and Premio Chiusura. As a three-year-old, he captured another seven races, including the Premio Parioli (the Italian 2,000 Guineas); the Gran Premio del Re (mile-and-a-half Italian derby); the Gran Premio del Impero; the Gran Premio d'Italia; the Gran Premio di Milano; and the Grand Prix de Paris. The latter race had been won by Nearco's maternal grandsire, Spearmint, thirty-two years before. The dark colt retired undefeated, having won over distances ranging from six furlongs to one mile and seven furlongs.

After his racing career, Nearco was purchased by Martin Benson for £60,000 and sent to Beech House Stud at Newmarket.

Nearco gained a reputation as a difficult stallion yet built a close relationship with stud groom Ernest "Ernie" Lee. During World War II,

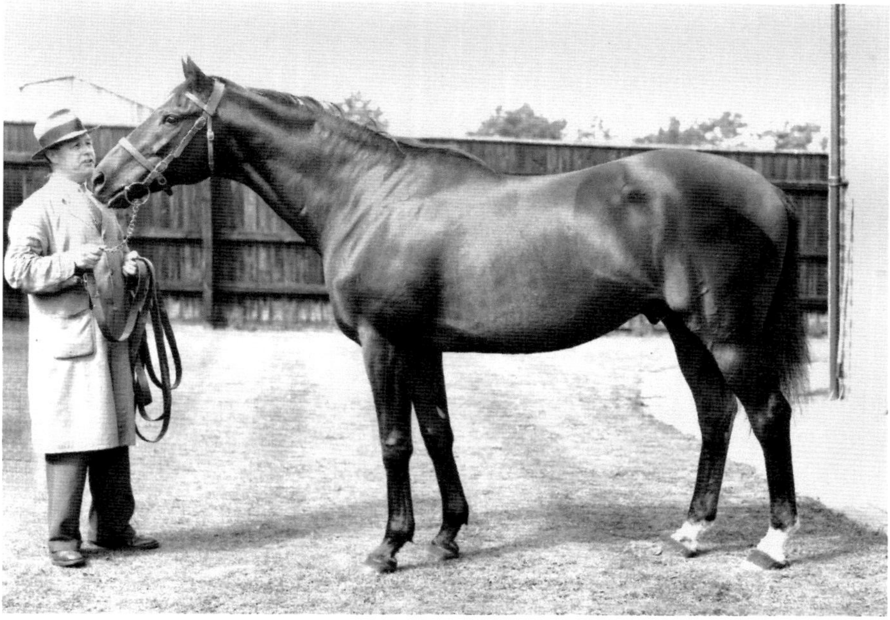

Nasrullah's sire Nearco at Beech House Stud, Newmarket, 1955. *Courtesy of Segas Picture Gallery (http://www.segaspicturegallery.co.uk/index.asp).*

the two first sought refuge in the south of Wales, and then, upon their return to Beech House, a paddock with a specially constructed cement underground bomb shelter was built for the valuable stallion. Lee trained Nearco to walk underground at the first wail of the air raid siren and then wait quietly with Lee until the "all clear" was given.

Nearco was humanely destroyed in 1957 at the age of twenty-two, after suffering from a cancerous hip and digestive problems. Apparently, Lee was so devastated by the stallion's death that he left his job at the stud and never worked with horses again.

Nearco was twice champion sire and was among the ten leading sires for sixteen consecutive years. His classic winners included colts Dante (winner of the Epsom Derby); Nimbus (2,000 Guineas and the Epsom Derby); Sayajirao (Irish Derby and St. Leger); Hafiz II (Champion Stakes); fillies Masaka (English and Irish Oaks and 1,000 Guineas), Neasham Belle (Oaks) and Noory (Irish Oaks); and Neolight (Coronation Stakes). His success as a racehorse and stud is not surprising if we are to examine his own bloodline more closely.

Pharos

Nearco's sire, Pharos, was bred by Edward Stanley, the seventeenth Earl of Derby.

> *The Earl of Derby's Stanley House Stud was the most important breeding and racing operation of its time, in the first half of the Twentieth Century. Two of the most important products of that stud were the full brothers Pharos and Fairway, bred on the famed Pharos/Chaucer nick. While Fairway was the more gifted and classic of the pair, the speedier Pharos proved the more influential stallion, and his male line became one of the most significant in the last fifty years.*[14]

Pharos's dam, Scapa Flow, which was bred and owned by Lord Derby, was sired by Chaucer, twice leading sire of broodmares and sire of Selene,

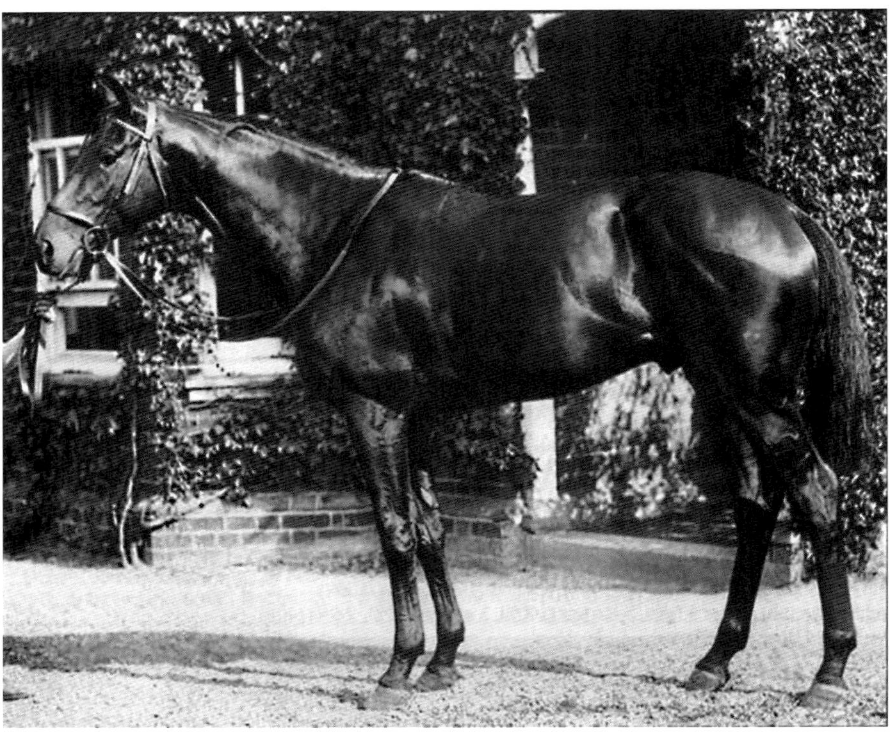

Pharos, Nasrullah's paternal grand-sire. *Courtesy of Wikipedia Commons.*

dam of the great Hyperion, which, despite being deaf, dominated the racing scene with wins in the derby and St. Leger Stakes and also went on to become a six-time leading sire and four-time leading broodmare sire. Scapa Flow's maternal grandsire was Ascot Gold Cup–winner Love Wisely. Scapa Flow's other foals included Pharos's classic-winning full siblings Fairway (St. Leger) and Fair Isle (1,000 Guineas), as well as stakes winner Fara.

Pharos won six of ten starts as a two-year-old and placed second in the 1923 Epsom Derby, winning three other stakes races as a three-year-old. He continued to race and win as a four- and five-year-old and retired having won fourteen races in thirty career starts.

At stud, Pharos first stood at Woodland Stud at Newmarket. He was leading sire in England in 1931 and again led the sire list in 1939 in France, where he relocated in 1928. Aside from Nearco, he is also known as the sire of the undefeated Pharis, derby-winner Cameronian and St. Leger–winner Firdaussi.

Phalaris

Pharos's own sire, Phalaris, was leading sire, leading sire of two-year-olds and leading broodmare sire in Great Britain and Ireland on numerous occasions.[15] He is arguably the most influential sire of the twentieth century. Phalaris's dam, Bromus (by derby-winner Sainfoin) was purchased by the seventeenth Earl of Derby on the advice of his racing manager and trainer, George Lambton.

On the track, Phalaris was a stakes winner through the ages of two to five and became known as a great weight carrier, carrying more than ten stone (63.5 kilograms) on three occasions. He retired having started twenty-four times and having won sixteen races. Phalaris's best distance was between five and six furlongs, and he had trouble staying the distance in longer races when the pace was fast.

Of the horse, Lambton wrote, "He had absolutely first-class speed, an excellent constitution, and was up to very high weights. Furthermore, he had very good action and was as true as steel as far as he could go."[16]

At stud, Phalaris sired five classic winners. His best offspring included Pharos and his full brother, Fairway, winner of the St. Leger and Eclipse Stakes; derby-winner Manna; and Sickle, twice leading sire in the United States.

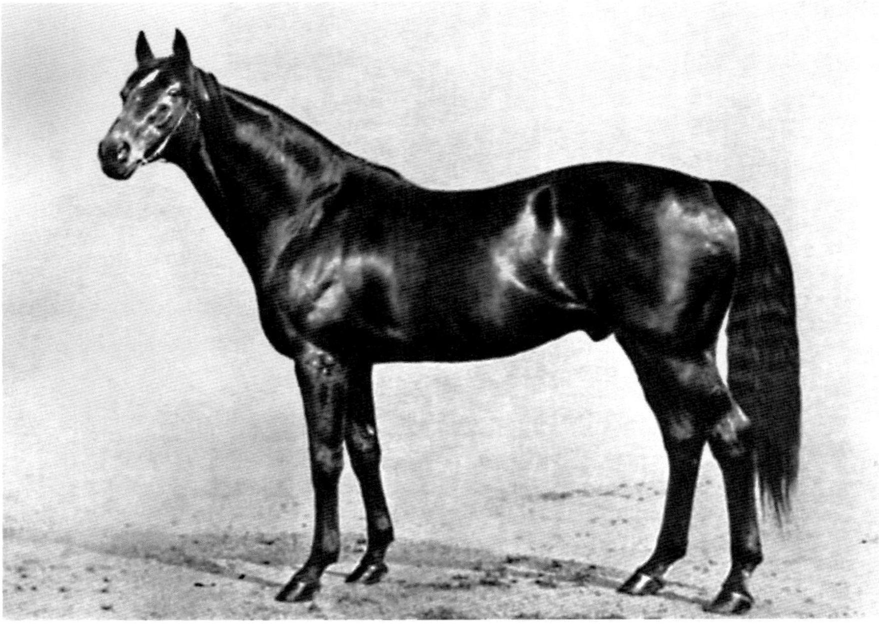

Phalaris, one of the most influential sires of the twentieth century. *Courtesy of Wikipedia Commons.*

Polymelus

Phalaris's sire, Polymelus, was bred by the Earl of Crewe. He was sired by Ascot Gold Cup–winner Cyllene,[17] a champion sire in England and Argentina by 2,000 Guineas–winner Bona Vista, out of the mare Arcadia (her sire was the Cups Triple Crown–winner Isonomy). A good runner, Polymelus placed in a number of stakes and classics at three, including the St. Leger, but he was never considered to be in the same class as the classic winners of his generation. At stud, however, he became the most successful British stallion of his era, leading the sire list five times and siring five classic winners: Black Jester (St. Leger); Pommern (Triple Crown); Fifinella (derby and Oaks); Cinna (1,000 Guineas); and Humorist (derby).

Chapter 3

RACING AT NEWMARKET

Not much is known of Nasrullah's early years at Sheshoon Stud. As a two-year-old, he was sent to England to the Aga Khan's trainer, Frank Butters, to be readied for his racing career. Butters, born in 1878, was a native of Austria and began his racing career in his home country before moving to Italy and arriving in England in 1926 to train for Lord Derby. In his first season at Newmarket, he won the Oaks with Beam and was named champion trainer at year's end. The next year, he was again champion trainer, taking the Oaks, this time with Toboggan, and the St. Leger with Fairway. In all, Butters was crowned champion trainer eight times. In 1931, he began training for the Aga Khan, who described him as the "greatest trainer of all."[18]

During his career, Butters trained a total of fifteen English classics winners, many of which raced for the Aga Khan and brought success to Sheshoon Stud, including British Triple Crown–winner Bahram and Epsom Derby–winner Mahmoud. The Aga Khan was quick to note of the trainer, however, that "even more wonderful than his success with great horses was his way with quite moderate ones. He had a wonderful knack of getting out of any horse the very best that horse could do."[19]

Butters certainly had his work cut out for him with Nasrullah. On the track, Nasrullah quickly gained a reputation for being an unpredictable runner, and at times his behavior was absolutely obstinate. He was known for his tendency to "prop," or slow down, once gaining the lead; his son Nashua also exhibited this same undesirable trait at times. It has been

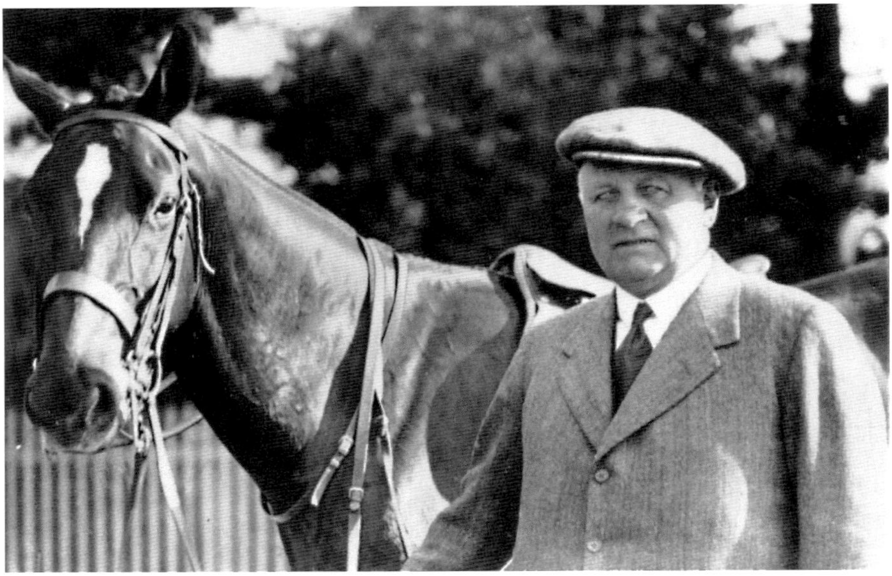

Frank Butters, Nasrullah's trainer. *Courtesy of the author.*

speculated that this behavior, exhibited by other British-trained colts, might have resulted from the common practice of having young colts follow a lead horse to help them settle down and learn not to pull while being worked. As Hewitt, author of *Sire Lines*, explains:

> *These colts may have become so accustomed to following other horses that they become acutely uncomfortable when asked to go to the front on their own, and hence sometimes pull themselves up.*[20]

It is unclear whether Nasrullah's attitude on the racecourse was a result of his own willful temperament, the restricted racing conditions under which he trained during the war, the training methods fashionable in Britain at the time or perhaps some combination of the three. Regardless, of the horse, Timothy Capps had the following to say:

> *Nasrullah had a wonderful physique and a horrid temperament, one that probably kept him from being the great racehorse his pedigree suggested he should be…* [he] *had superior speed but only an intermittent desire to use it.*[21]

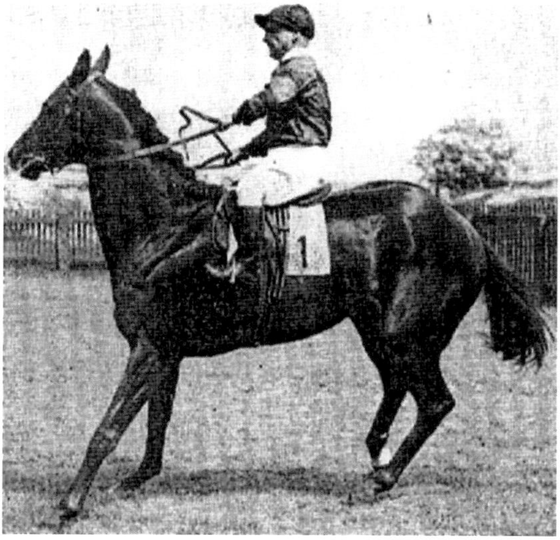

Appearing in the book *The Best Horses of 1943*, the original caption for this photo read, "Nasrullah impersonating a mule." *Courtesy of the Portway Press.*

Once, when Nasrullah refused to step onto the track at Newmarket, Frank Butters was later heard to comment, "I've had some brilliant Nearcos, but sometimes I never want to see another in the yard."[22] Indeed, Nasrullah was once described as having a "diamond-studded pedigree and the disposition of a paranoid mule."[23]

Jockey Gordon Richards described at length the trouble the horse caused:

> *We had terrible times with him in the early morning work. He would frequently hold the whole string up, and we had to dodge between the banks, and get up to all sorts of tricks with him, in order to get him to work at all.*[24]

Richards rode Nasrullah in six of his ten races. He was one of the leading flat-racing riders in the mid-twentieth century, crowned champion jockey an astonishing twenty-six times between 1925 and 1953, and he set a record of 4,870 winners in Britain that remains unbeaten to this day. This also stood as a world record until 1957, when English-born American jockey Johnny Longden became the first to ride 5,000 winners. At age forty-nine, Richards was knighted by Queen Elizabeth II in Her Majesty's Coronation Year 1953. He remains the only jockey to have received this honor.

Richards was born in 1904 and grew up in Oaken Gates, Shropshire, the son of a miner and raised as a Methodist. While his parents were not horse

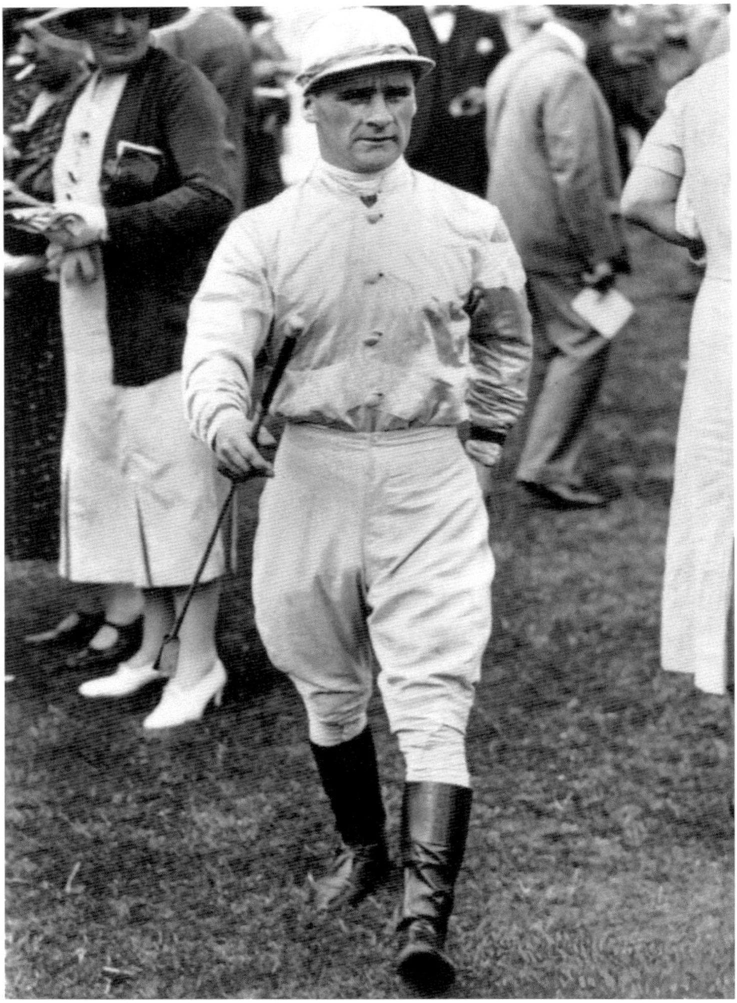

Sir Gordon Richards, Nasrullah's jockey. *Courtesy of the author.*

people, they built stables and kept ponies on the four-acre family property. As young boys, Gordon and his brothers would hitch the ponies to a trap and pick up passengers at the train station, driving them to their destinations and giving the money they earned to their mother to pay for their and the ponies' keep. After leaving school at thirteen, Richards worked as a junior clerk before seeking a job as a stable boy with Martin Hartigan at Foxhill when he was fifteen.

Richards began his racing career as an apprentice at Foxhill and was heavily influenced by top jockey Steve Donoghue. He rode his first winner, Gay Lord, in an apprentice race at Leicester in 1921. In 1925, he was retained by Thomas Hogg, for whom he rode his first classic winner, Rose of England, in the 1930 Oaks. In 1932, Richards began riding for six-time champion trainer Fred Darling, gaining the mount on classic-winners Pasch (2,000 Guineas); Sun Chariot (filly Triple Crown); and champion two-year-old Tudor Minstrel. In 1933, he rode 259 winners, beating Fred Archer's 1885 record of 246; that year, he also rode twelve consecutive winners. In 1943, while still racing for Darling, Richards was approached by the Aga Khan and trainer Frank Butters, who asked him for a second retainer, to which he eagerly agreed.

While Richards acknowledged that Nasrullah "was very very difficult to ride,"[25] in his autobiography, *My Story*, Richards recalled, "I have never ridden a better mover than Nasrullah, and I am sure his outbreaks of temperament were very largely due to the zoned racing of the war."[26] Horse racing was one of the only sports that continued during both world wars. During World War II, racing was restricted; only five racecourses remained open in Britain in 1942, the year Nasrullah began his campaign as a two-year-old. During this time, a horse could not compete outside its zoned area, and Nasrullah was confined for training and racing purposes to Newmarket, in Suffolk, which remained the only British racecourse to host racing during the two world wars.

Newmarket has a long and noble history. The area where the track is now situated was given as a dowry in 1200 to Sir Richard de Argentein when he married Cassandra, daughter of Robert de Insula, Lord of the Manor of Exning. The heath where the track was built was discovered in 1601 by King James I while he was hunting; he saw the spot as a potential leisure venue for his court. The first recorded race at Newmarket was on March 18, 1622. It was a match race between a horse owned by Lord Salisbury and a horse belonging to the Marquis of Buckingham. Buckingham's horse won, claiming a then whopping £1,000. King Charles II, known as the "Merry Monarch," moved his royal court to Newmarket each year. An avid equestrian, he rode over what is now known as the July Course. In 1666, the Town Plate, the first race to be run under written rules, was staged; it is still run each October and is the oldest surviving horse race.

Nasrullah made his racing debut in the Wilburton Stakes, a five-furlong affair, on June 12, 1942, Oaks Day at Newmarket.[27] Starting at odds of eleven to eight in a field of eight, he was ridden by jockey Doug Smith.

Unable to close in time for the wire, he finished third on a firm track, beaten by one and a half lengths by his half-sister, Nearly (Nearco–Lost Soul), which carried the top weight of nine stone, four pounds, and to which he had conceded nine pounds. The colt Harroway was in second.

Nasrullah started in the five-furlong Coventry Stakes on June 30 at odds of seven to four amongst a field of seven, all carrying nine stone. Normally run at Ascot, the race was transferred during the war to Newmarket. This time, Smith was replaced by jockey Richards. After a slow start, Nasrullah eventually gained speed and sped past the leaders, winning by one and a half lengths and defeating the next year's derby winner, Straight Deal, as well as the colt Victory Torch.

On July 28, with Doug Smith back in the saddle, Nasrullah won the six-furlong Great Bradley Stakes by four lengths as the odds-on favorite at six to one hundred and carrying 131 pounds. His only opponent was Lord Derby's filly Feriol, to which he conceded eleven pounds.

On August 26, Nasrullah was in a field of eight for the Middle Park Stakes at six furlongs, again the odds-on favorite (eleven to eight) and again piloted by Smith. The colts all carried nine stone with a three-pound fillies' allowance. The start of the race was delayed when Straight Deal threw his rider and ran off into a nearby field. Nasrullah made his move too late and, despite a game effort, lost by a neck to the filly Ribbon. Nearly finished third.

In the two-year-old Experimental Free Handicap for 1942,[28] Nasrullah was rated the best two-year-old colt at 132 pounds, but 1 pound below Lady Sybil, a Nearco filly, which was assigned 133 pounds. V.R. Orchard, writing in *The Bloodstock Breeders' Review*, was appreciative in his physical description of the maturing colt:

> *Whatever else may be said of Nasrullah, it is clear that he is a colt of character. In appearance he is a rich bay of commanding proportions. His quarters are immensely powerful, and any good judge of a horse would put down this fine-looking colt as near perfect as possible if considering him apart from his racecourse performance.*[29]

As a three-year-old, Nasrullah won three of his six races. Carrying nine stone, two pounds in the one-mile Chatteris Stakes on May 5, he was reunited with Richards and made the even-money favorite. Despite giving up once he gained the lead, he managed to get the win, "coasting past the winning post,"[30] ahead by half a length over Response and Herald.

His antics prior to the start of the race were well noted, however. According to British handicapper Phil Bull in *The Best Horses of 1943*, Nasrullah put on quite the display:

> *He refused to leave the paddock, he refused to break into a trot, he refused to respond to the brandishments of the friendly hack sent out on the course to kid him; he refused to do anything except behave like a spoilt child. And all the while Gordon sat on his back, flapping the reins at him with a patience that would have done credit to Job. Could the catcalls and cries of derision which greeted this unthoroughbred-like behaviour have been heard by Nearco across at the Beech House Stud, and could their origin have been explained to him, it might have had a serious effect on his fertility.*[31]

In the one-mile 2,000 Guineas on May 25, carrying nine stone, Nasrullah was the thirteen-to-eight favorite. This time he was outfitted with blinkers, considered to be the "rogue's badge" in English racing at the time, despite their common use in America. Unfortunately, this did not seem to improve his focus any. Richards held him back behind the leaders and then gave him his head. After passing the pack, he stopped trying an eighth-mile from home, and despite the jockey's attempts to coax him back into competition, he slipped back to finish fourth behind Kingsway, who got a head over Pink Flower and Way In. Richards, in remembering the race, expressed his disappointment in the horse:

> *He ran a very unsatisfactory big race. First he looked like winning, then he would not go on, then he suddenly decided to gallop again, and just as soon decided not to. It was exasperating for me, because he should have won by lengths.*[32]

Nasrullah followed this up with a repeat performance in the mile-and-a-half Epsom Derby on June 19. Again, he carried nine stone and started at odds of nine to one, with Richards once again in the irons. After going to the front, he dropped the bit and was beaten by Straight Deal and his stable companion, Umiddad. In his memoir, the Aga Khan, who was exiled to Switzerland during the war, recalled that he received word of his horse's performance in the race via telegram.[33]

After a two-month break, Nasrullah managed to win the mile-and-a-quarter Cavenham Stakes on August 18, carrying nine stone, four pounds and the favorite at odds of one to four. Though his efforts were minimal—despite

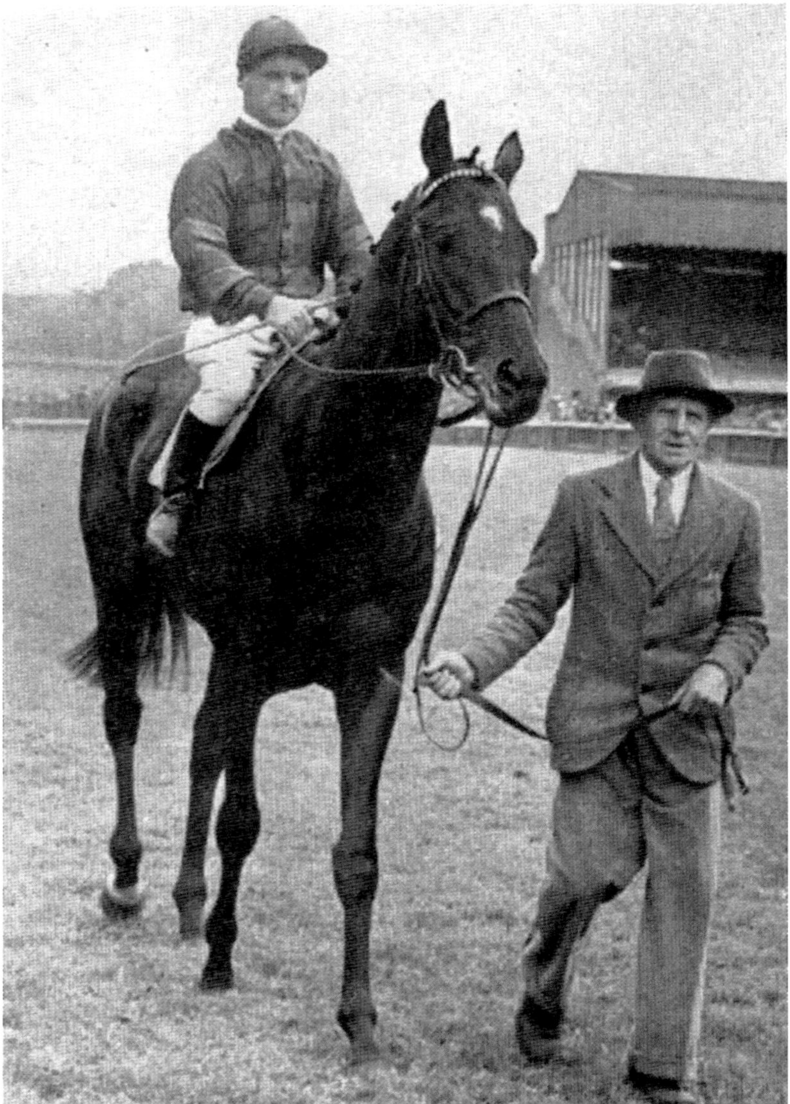

Nasrullah in the Epsom Derby parade with Richards and Butters. *Courtesy of the Portway Press.*

Richards urging him on once they gained the lead—he was ahead at the wire by two lengths over Triumvir.

Ridden by Michael Beary in the St. Leger[34] (mile and three-quarters) on September 18, at odds of ten to one (on the fourth line of betting), Nasrullah did not take the lead at any point and finished sixth behind the fillies Herringbone and Ribbon, with Straight Deal in third. Persian Gulf was fourth, followed by the filly Tropical Sun back in fifth. It was speculated that in this instance, Nasrullah did not have the stamina and was simply unable to stay the distance.

Nasrullah's last start came in the Champion Stakes, a ten-furlong (mile and a quarter) race on October 5, in which all runners carried 8 stone, 6 pounds, except Lady Electra, which carried 8 stone, 11 pounds. Nasrullah was made second choice at one hundred to thirty to the three-to-one favorite, Kingsway. Richards held him back until very late in the race and then unleashed him. Nasrullah took the lead and beat Kingsway, winner of the 2,000 Guineas, by a length, with Ummidad back in third. On the three-year-old Free Handicap, Nasrullah received 132 pounds, 1 pound less than Straight Deal, and the same as Herringbone, winner of the 1,000 Guineas and St. Leger.

Nasrullah retired after a two-year career of five wins from ten starts and earnings of $15,217.35.[35] He was evaluated as being "a good colt especially brilliant at middle distances,"[36] and Phil Bull concluded his write-up of the horse's three-year-old campaign with the following remarks:

> *You've only got to look at his racefigures, to see that he was a racehorse of exceptional ability. Personally, I have to announce that I still persist, with a pig-headed obstinacy equal only to that of Nasrullah himself, in the firm opinion that (up to 1½ miles), the son of Nearco was head and shoulders above his contemporaries in ability. Yes, I know that other qualities go to the making of a great racehorse—courage, determination, the will to race, tractability and so forth—I grant all that; my confession of belief in Nasrullah is confined to his ability. To say that he was a model thoroughbred would be tantamount to a libel on the breed; to say that he was a great racehorse, would be to ignore his failings; but to say that he could have been a great racehorse, if he had cared to, is quite a permissible expression of opinion. His photograph shows a good head with small ears; a strong neck, well developed withers and excellent depth from withers to brisket; a fair shoulder and a fine, well ribbed-up body; plenty of muscle over the humerus* [sic], *and full, well developed gaskins, strong straight hocks and pretty good*

legs generally. A high class, quality colt, possessing also range and size. He is more impressive in the flesh than in this photograph. None of his rivals of last season could beat him in appearance and none possessed a more delightful action or more space devouring stride when at full stretch.[37]

Chapter 4

OFF TO STUD

Before he even entered a breeding shed, Nasrullah's success as a sire could be said to have been prophesized by Phil Bull, who wrote the following in 1943:

> *If conformation and innate ability count for anything he may make the name for himself as a stallion which his unfortunate temperament prevented his making for himself as a racehorse.*[38]

Fully mature, Nasrullah was a strong and solid horse, standing sixteen hands, one and a half inches. At the conclusion of his three-year-old season, he was sent to stud at Barton Stud in Suffolk, England. Originally purchased by brothers Huttleston and Henry Broughton in 1925, the property was leased to the Aga Khan during World War II. Under the management of Major Fleming, the stallion initially stood for a fee of 198 sovereigns.

From Nasrullah's first crop, which arrived in 1945, came eventual Hall of Famer Noor, which defeated the mighty Citation four consecutive times; Irish Derby–winner Nathoo; and stakes-winner Golestan. The story of Noor and Nathoo is an interesting one. After several victories as a three-year-old, the gray colt Nathoo was sent by his owner and breeder, the Aga Khan, to race in the United States, where he faced Citation in the 1948 Empire City Gold Cup. Nathoo was unplaced behind the Triple Crown–winner and Belmont Stakes–victor Phalanx. Learning from Canadian Charles Spencer that the colt was for sale, Charles Howard, owner of the famed Seabiscuit, indicated his interest.[39]

Turning to Californian Rex Ellsworth for his opinion of the horse, Howard was advised to consider Nathoo's half-brother Noor, still in England, which had impressed Ellsworth with his third-place finish in the Epsom Derby, and Ellsworth felt he was the better horse. Howard ended up buying both Noor and Nathoo in a package for an undisclosed price. Nathoo never raced in America due to a bowed tendon and was unsuccessful at stud, while Noor would go on to defeat two Triple Crown winners and win an Eclipse award as champion male horse.

Also in 1945, Nasrullah was sold by the Aga Khan for £15,000 ($50,000) to Irish bloodstock dealer Bert Kerr and Dr. Gerald McElligott. Within a week, he was resold to Irish breeder Joseph McGrath for £20,000 ($76,000). In letting the stallion go, Kentucky breeder Dr. Eslie Asbury said, "It may have been the only mistake the Aga Khan ever made."[40]

Nasrullah was sent off to McGrath's Brownstone Stud on the Curragh of Kildare, Ireland, where he stood for five seasons. The property, which dates back to the late eighteenth century, was, at the time of this writing, offered for sale. Standing for a fee of 250 guineas, Nasrullah's studbook was full for 1946, the year his second crop of foals arrived.

Musidora was undoubtedly the star of the 1946 crop. The bay mare was bred in Ireland by F.F. Tuthill, who bought her dam, Painted Vale (sired by English Triple Crown–winner Gainsborough), for 220 guineas. As a yearling, Musidora was purchased at the Doncaster sales for 4,700 guineas by shipping magnate Norman Donaldson and sent to Yorkshire to be trained by Charles Elsey at his Highfield Stables. Musidora raced six times as a two-year-old, winning once and never running out of the money.

Under the guidance of Australian jockey Edgar Clive Britt, Musidora won two classics as a three-year-old, the 1,000 Guineas and Epsom Oaks Stakes, defeating future Prix de l'Arc de Triomphe winner Coronation in the latter. Her winning time of 2:40.0 was two seconds faster than that recorded for Nimbus in the Epsom Derby over the same course and distance two days later.[41] She also won the Roseberry Stakes, came in second in the Malton Stakes and was unplaced in the Yorkshire Oaks, the St. Leger and Doonside Cup.

A prep race for the Oaks, the Musidora Stakes, run annually at York since 1961, is named in her honor. As a broodmare, she produced three winners, including the race mare Wishful Thinking.

Other noteworthy offspring from the 1946 bunch included stakes-winner Callander; Nasr-ed-din, two-time leading sire in Chile; American steeplechaser El Arabi; and broodmare Ballynash, which produced six winners, including Mourne and Petite Sanguenay.

Stakes-winner Tambara (Newmarket Rous Memorial Stakes and Royal Coronation Stakes) and stakes-placed Catchit (runner-up in the 1,000 Guineas) were noteworthy foals of 1947.

Foals from Nasrullah's 1948 crop included classic-winner Belle of All, out of the mare Village Beauty. Bred in Ireland by the Earl of Dunraven, the bay filly was sent to the Doncaster sales as a yearling, where she was sold for 8,000 guineas to Henry Thanet Tufton. Belle of All went to trainer Norman Bertie's Bedford Cottage stable at Newmarket.[42] She won her debut race, the National Breeders' Produce Stakes, a five-furlong affair, defeating eventual Hollywood Gold Cup–winner Royal Serenade, which ran second, and half-brother Grey Sovereign, which finished third. That fall, with Gordon Richards in the saddle, she won the Cheveley Park Stakes, Britain's most prestigious race for two-year-olds.[43] At year's end, she was ranked equal with Gamble in Gold as the best filly in the Free Handicap, a rating of the best two- and three-year-olds.[44]

As a three-year-old, Belle of All won the 1,000 Guineas in smashing form as the four-to-one favorite, beating seventeen other fillies.[45] She was unable to stay the mile-and-a-half distance of the Oaks, however, finishing third to Neasham Belle and Chinese Firecracker.[46] She returned to the winner's circle with a victory in the one-mile Coronation Stakes, winning by five lengths,[47] but finished her racing career with a disappointing fourteenth-place performance in the inaugural running of the mile-and-a-half King George VI and Queen Elizabeth Stakes.[48] Belle of All produced several winners at stud, and her son Pendragon became a successful National Hunt stallion.[49]

Zucchero was another outstanding foal of 1948. *Zucchero* is the Italian word for "sugar," and if it was meant as a term of endearment, the fiery colt did not live up to expectations with regards to temperament, instead following in his father's footsteps and becoming known as a bit of a rogue. Phil Bull, in *Racehorses of 1953*, described him as

> *one of the most exasperatingly unpredictable racehorses of recent times, prepared to race only when he felt so inclined, and absolutely immovable in his determination not to race when he didn't feel like it. On his day he was a brilliant racehorse: on his off-days he was impossible!*[50]

Out of the mare Castagnola, Zucchero was bred by Captain Spencer-Freeman and owned by George Rolls. Ken Cundell, who often had to

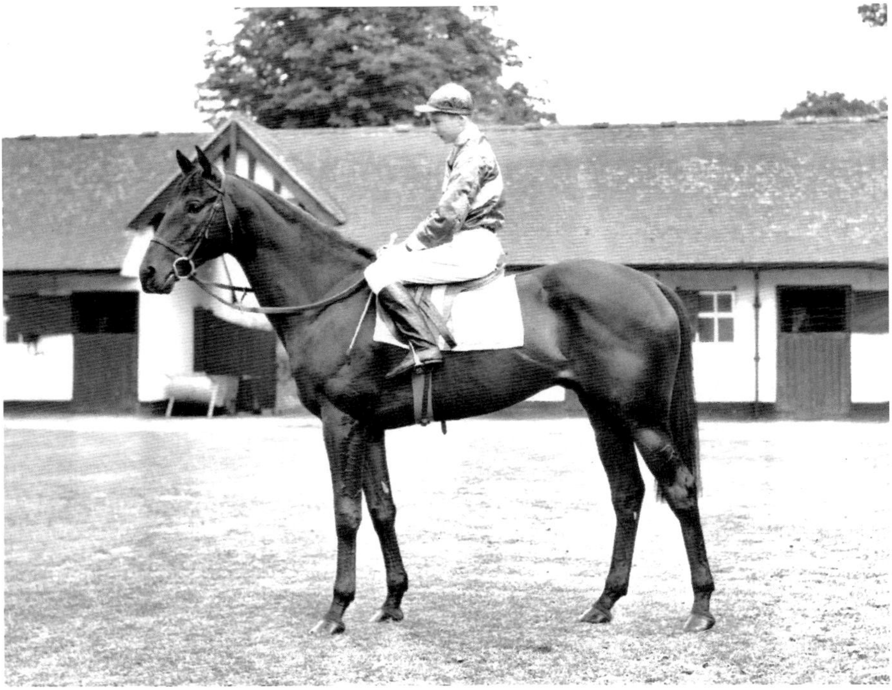

Zucchero with Lester Piggott up, 1953. *Courtesy of Segas Picture Gallery (http://www.segaspicturegallery.co.uk/index.asp).*

accompany the horse to the start of his races, leading him to the starting tape, first trained him. As a three-year-old, the brown colt won the Blue Riband Trial and Commonwealth Stakes. He was Lester Piggott's first mount in the Epsom Derby but did not put forth much of an effort, finishing a dismal thirteenth. He did manage to finish second in the first running of the King George VI and Elizabeth Stakes.

In 1952, after being transferred to trainer Bill Payne, Zucchero won the Prince of Wales Stakes and finished third in the inaugural Washington, D.C. International.[51] As a five-year-old, his wins included the Coronation Cup and the Rose of York Stakes (for the second time). He was voted champion older horse for that year. Zucchero was not a successful sire, despite attempts at stud in both England and the United States, and was sent to Japan in 1961.

Nasrullah's 1948 crop also included stakes-winners Cassydora and Malka's Boy, as well as Grey Sovereign, winner of the Richmond Stakes and eventual champion British sire of two-year-olds in 1958 and 1961 and champion French sire of two-year-olds in 1967. This was also the year Noor ran third in the Epsom Derby and was imported to the United States.

Foals of 1949 included stakes-winners Orgoglio and Rego. Out of the mare Orienne, Orgoglio was a handsome chestnut colt. He won several stakes as a two-year-old, including the Champagne Stakes, and continued to race and win at three, four and five. Sent to Arundel stud in Australia at age six, he went on to sire twenty-one stakes winners. Rego, a brown colt out of the mare Missy Suntan, won the eight-furlong Epsom Plate as a three-year-old, and at four Rego captured the ten-furlong Prince of Wales Stakes. Like his half-brother, he was sent to Australia for stud duties, where he sired the champion Baguette (undefeated at two and sire of eighteen stakes winners) and Wiggle, a stakes winner in Australia and the United States.

The first of two sons of Nasrullah to be named Nashua was also a foal of 1949. The "original" Nashua raced for the Aga Khan and was a winner of the Pretty Polly Stakes and Irish 1,000 Guineas. Indian Hemp was also a winner in both England and the United States and sired twenty-two stakes winners, including 1974 North American leading sire T.V. Lark.

Nearula was a foal of 1950. Out of the mare Respite, the handsome colt was owned by William Humble and trained by Captain Charles Elsey. As a two-year-old, Nearula won the Middle Park Stakes and was the high-weighted juvenile colt; at three, he triumphed in the 2,000 Guineas, the Thirsk Classic Trial and the St. James Palace Stakes, and he managed third-place finishes in the Knight's Royal Stakes and the Scarborough Stakes. At four, Nearula won two stakes races, the Burwell and March Stakes, and finished third in the Coronation Cup. The bay horse stood only three seasons at stud before his untimely death of a ruptured blood vessel. Captain Kidd; Red Gauntlet; Primula, Italian stakes winner; and Kythnos, winner of the Irish 2,000 Guineas, were among his winners.

Nasrullah's seventh and final crop to be born in Ireland included classic-winner Never Say Die. Out of the War Admiral mare Singing Grass, this chestnut colt was born at Jonabell Farm and was bred and owned by Robert Sterling Clark, heir to the Singer Sewing Machine fortune and an American art collector. Conceived in Ireland, foaled in Kentucky and sent to England as a yearling, Never Say Die was the first American-bred and American-owned winner of the Epsom Derby, giving eighteen-year-old jockey Lester Piggott his first derby win. He demonstrated his ability to master classic distances with a twelve-length victory in the St. Leger, the biggest winning margin recorded in the race's history.

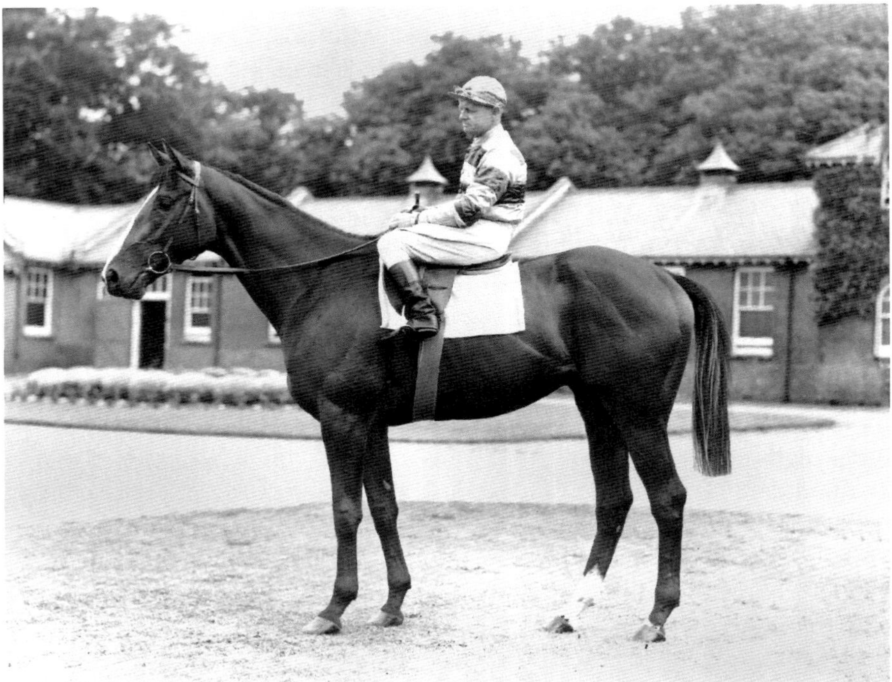

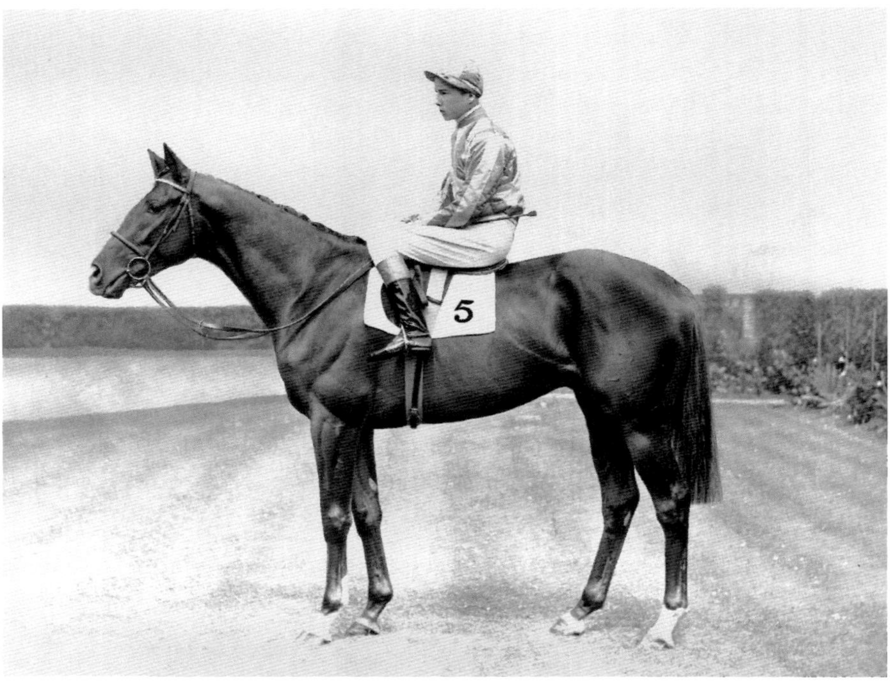
5

Following his retirement from racing, Never Say Die was donated to the National Stud. He was leading sire in England in 1962 and was also a leading sire of broodmares. Despite gaining a reputation for siring horses with soft bones, his son Larkspur captured the Epsom Derby and his daughter Never Too Late won the Oaks and 1,000 Guineas. His other winners include St. Leger runner-up Die Hard and Sostenuto.

The brilliant broodmare Dangerous Dame also came from Nasrullah's last Irish-conceived crop. The chestnut mare was bred in England by David Vard. She was out of Lady Kells, champion two-year-old filly in Ireland in 1946, which was by the Hyperion stallion His Highness and out of the mare Anyway, which also foaled the winner Solonaway. Harry F. Guggenheim bought Dangerous Dame as a yearling and imported her to the United States, where she won two of ten starts and earned $9,200.[52] As a broodmare, she is best known for producing the stakes-winning fillies Hidden Talent and Heavenly Body, which were also her most productive daughters. Her other foals included His Lady Fair, an unraced mare by four-time Eclipse Award–winner and Hall of Famer Tom Fool, which produced the stakes-winner Diplomette.

Other noteworthy progeny of 1951 included Princely Gift, stakes winner in England and a leading sire of two-year-olds (he was third on the 1966 list and second in 1967), and handicappers Mister Gus and Coronation Boy.

Nordlys would go on to establish a legacy of successful jumpers, siring multiple international-level show jumpers, including Ambassador, a stunning dapple gray gelding with a flowing white tail, which in 1972 became the first Irish-bred show jumper to win an Olympic individual gold medal in the sport. He also helped the Italians to a team bronze at those games. Nordlys also sired Kwept, which helped team Germany take the silver medal at the 1976 Olympics in Montreal.

While Nasrullah met with success as a sire in Great Britain and Ireland, he did not leave as lasting an impression there as he had the potential to do or as he would in the United States. This might have been due, in part, to the fact that he covered only a small number of mares each year, and these were not always quality broodmares or of the strongest bloodlines. And then there was his temperament.

Opposite, top: Nearula with jockey E. Britt, 1953. *Courtesy of Segas Picture Gallery (http://www.segaspicturegallery.co.uk/index.asp).*

Opposite, bottom: Never Say Die with jockey Lester Piggott, 1954. *Courtesy of Segas Picture Gallery (http://www.segaspicturegallery.co.uk/index.asp).*

Initially, breeders might have been reluctant to send their mares to the stallion, fearing that his fiery spirit would be passed on and result in colts and fillies that were difficult to train and inconsistent runners on the racetrack. Gordon Richards recalls that Michael Collins, who trained for Joe McGrath, came to him and asked his opinion on sending mares to the horse. Richards responded favorably, telling him that he believed the stallion's outbursts were the result of being confined at Newmarket.[53]

Nasrullah made the list of leading stallions on numerous occasions, ranking eighth on the 1948 sire list, third in 1949 and seventh in 1950 before heading the list in 1951. In 1952, he was again in third place and second in 1953 and 1954. But it would be in America that Nasrullah would establish his most lasting legacy.

Chapter 5

AMERICA CALLS

While Nasrullah might have been the most influential Thoroughbred stallion imported to North America, his legacy was preceded by a long history of non-native (foreign-bred) leading sires. This is no surprise given that the breed was first established overseas, with the importation of the three foundation sires—the Byerley Turk, the Darley Arabian and the Godolphin Arabian—to England from the Middle East in the late seventeenth and early eighteenth centuries.

The first Thoroughbred to have been imported to North America was reportedly Bulle Rock, a son of the Darley Arabian and out of a Byerley Turk mare, which arrived in Virginia in 1730. Due to poor and inaccurate record keeping, however, information on the horse is sparse; the next stallion for which there are records is Crab, imported in 1746.[54]

Diomed, first winner of the Epsom Derby, was imported to Virginia in 1799. The list of leading sires in the nineteenth century was dominated by native-bred horses—beginning in 1860 with Revenue,[55] Virginia bred out of the British stallion Trustee—and controlled for almost two decades by Kentucky-bred Lexington, a great-great-grandson of Diomed, which led the sire list a whopping sixteen times and for fourteen consecutive years between 1861 and 1874.

Before 1930, only five imported stallions led the sire list in the twentieth century: Ben Strome (1903); Meddler (1904, 1906); Star Shoot (1911,1912, 1916, 1917, 1919); McGee (1922); and Chicle (1929). A gradual shift began with the 1926 importation of Sir Gallahad III, which led a total of four times between 1930 and 1940 and shared this position over that decade with imports St. Germans, Sickle and Challenger II. Imported from France, the stallion sired three Kentucky Derby winners, including Triple Crown–winner Gallant Fox, which in turn sired Triple Crown–winner Omaha. Sir Gallahad's brother Bull Dog, which was imported in 1930 and led the sire list in 1943, sired five-time leader Bull Lea, which joins Star Shoot and Nasrullah as the only twentieth-century sires to lead five times.

Nasrullah's maternal grandsire Blenheim II was imported after the 1936 breeding season—the first of three derby winners to be sold by the Aga Khan III. The stallion was purchased for $225,000 to a syndicate headed by Arthur B. Hancock Sr.[56] Standing at Claiborne Farm, Blenheim led the sire list in 1941 and sired Triple Crown–winner Whirlaway.

Blenheim's son Mahmoud, winner of the 1936 Epsom Derby, as well as his close relative 1935 British Triple Crown–winner Bahram, was also sold by the Aga Khan III and imported in 1940. The former was sold for $84,000 and stood at stud for C.V. Whitney, leading the U.S. sire list in 1946, while the latter was sold for $160,000 and later sold to a stud farm in Argentina. Bahram had a profound impact on the Holstein and Westphalen breeds. The Aga Khan's sale of these great stallions has been met with some controversy over the years by the British breeding industry but defended as being necessary from a financial point of view. According to the Aga Khan Studs:

> *During the war years financial necessity was behind the Aga Khan's decision to sell his Derby winners rather than aspire to embellish his record. In 1940 he parted with Bahram and Mahmoud for a total of £60,000, a fraction of their value on an open market.*[57]

The Aga Khan still managed to be the leading breeder in England in 1941 and 1942; this was facilitated in part by the success of Nasrullah in his two-year-old campaign.

Heliopolis was Nasrullah's immediate predecessor as a leading sire in America. The son of Hyperion was imported from England and entered stud in Kentucky in 1941. He led the sire list in 1950 and 1954. The only two horses to interrupt Nasrullah's reign as leading sire between 1955 and 1962 were his fellow imported stable mates Princequillo and Ambiorix.

Princequillo was smuggled through German blockades as a weanling; his own sire, Prince Rose, was killed by artillery fire while standing at stud in France. Princequillo was imported to the United States in 1941 and claimed by Horatio Luro for $25,000 on behalf of Audrey Emory and Prince Dimitri

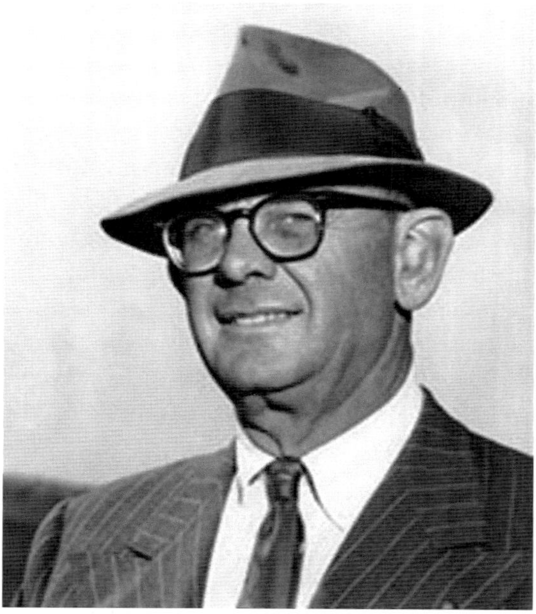

Arthur B. "Bull" Hancock Jr. *Courtesy of Stone Farm.*

Djordjadze, for whom he raced before being sent to Ellerslie Stud in 1944. He became leading U.S. sire in 1957 and 1958 and was also leading broodmare sire on eight occasions. Ambiorix was champion two-year-old in France and won two of four starts as a three-year-old before being snatched up by Hancock to stand the 1950 stud season at Claiborne. He led the sire list in 1961.

Nasrullah first caught the eye of Arthur B. "Bull" Hancock Jr. of the fabled Claiborne Farm in Paris, Kentucky, while the colt was still racing in England. Qualities he saw to his liking in Nasrullah were speed, early maturity, ability to race well in the classics and pedigree. Bull's goal was to win the Kentucky Derby, and he wanted a Nearco stallion. Rustom Sirdar was the only one in the United States at the time. Interestingly, it would be some time before the Nearco line was favored in Europe. In Bull's own words, "I picked him out because he was the best two-year-old of his year, was third in the Epsom Derby, and got one and one quarter miles very well."[58]

The Hancock family is renowned in American racing and breeding circles for its success as an independent breeder, and it has played a monumental role in the shaping of the Thoroughbred breed on an international scale. Claiborne Farm is in fact likely the most influential American breeding operation in history. Now in its third generation of family management, Claiborne continues its domination into the twenty-first century and in 2010 celebrated its one-hundred-year anniversary.

But while Claiborne came to be the preeminent Thoroughbred establishment in Kentucky in the second half of the twentieth century, the state had become known as "horse country" much earlier in its history. In the first half of the nineteenth century, horse racing was centered in the

East Coast states of New York and Maryland. While Kentucky's breeding industry was established in the late eighteenth century and flourished in the mid-nineteenth century—helped by the stud success of Lexington in the 1850s—racing didn't take off until the 1880s and 1890s. When gambling was made illegal in many states, multimillionaire gamblers, many of them horse people, flocked to Kentucky, where they set up premier breeding farms. Many of these people, as well as Kentucky's original settlers, came from Virginia, which also has a rich horse culture.

Bull Hancock's grandfather, Richard Johnson Hancock, had operated Ellerslie Stud in Albemarle County, Virginia, and had great racing success, winning the 1884 Preakness Stakes with Knight of Ellerslie. Hancock Sr. managed the Virginia property, along with Claiborne, until 1937, when Bull took over Ellerslie, and in 1946 the Virginia farm was sold to Robert Schlesinger. A 1954 Nasrullah colt was named in honor of the estate.

Today, Kentucky is known worldwide as the "horse capital of the world." In 2012, the Kentucky Equine Survey was conducted by the state's field office of the National Agricultural Statistics Service, with the support of the University of Kentucky College of Agriculture and the Kentucky Horse Council. It was found that Kentucky is home to 242,400 horses, the estimated value of which is $6.3 billion, and the total value of the state's equine industry is estimated at $23.4 billion. Thoroughbreds are the most prevalent breed in the state, numbering 54,000. In recent years, Kentucky-breds have dominated the sport of Thoroughbred racing, with more graded stakes winners than any other state, and they have set the standard in Thoroughbred sales.[59] Kentucky-bred stallions lead the national sire list more often than not, the majority of which descend from Nasrullah.

Arthur Hancock Sr., described by the *Bloodstock Breeders' Review* as "probably the most influential breeder in the history of the American turf,"[60] had acquired the property that was to become Claiborne, which spreads over three thousand acres in Paris, Kentucky, from the family of his wife, Nancy Tucker Clay. Claiborne began its breeding operation in 1910; among the farm's early stallions were Ambassador IV, grandsire of Jamestown; Omar Khayyam, winner of the 1917 Kentucky Derby; and Preakness-winner War Cloud.

While he was to later inherit Claiborne in 1957 from his father and namesake, Bull had begun managing the family farm much earlier, in 1949, when his father's health began to fail. Educated at Princeton University, where he played football and baseball and studied genetics, Bull served as a major in the United States Army Air Force during World War II. He was discharged and returned home in 1947 after his father suffered the first of several strokes. An avid horseman, Bull was a logical choice to lead the family business of breeding champion Thoroughbred racehorses. In a Claiborne customer's words, "He is the most knowledgeable horseman in the world today."[61]

By this time, Claiborne was in need of new blood in its broodmare band: Blenheim was twenty-three years old, and Sir Gallahad III had died the year before. Bull looked to out crossing and reconstructing the band of mares as strategies to revive Claiborne's stock and return the farm to its former glory. Claiborne's return to prominence in the postwar years after a slight slump owes much to the importation and success at stud of its leading stallions, Nasrullah in particular.

In making the decision to purchase Nasrullah, Bull followed his father's tradition of importing leading European stallions. Sir Gallahad III was the first of the Hancock's foreign acquisitions.[62] Hancock Sr. had also been part of a consortium that imported Blenheim II in 1936 and, upon his son's advice, in 1944 purchased the Irish-bred Princequillo,[63] which had been imported by his breeder, Laudy Lawrence, as a yearling. These purchases of overseas stallions catapulted Claiborne to the top of the American breeding and racing industry and helped put Kentucky on the map as a premier location for the raising of quality stock.

Bull was also following in the footsteps of both his grandfather and father as a skilled horse dealer. All three men had set their sights on a "big horse" and were willing to make significant financial investments in order to acquire them. Richard Hancock chased a stallion named Eolus for six years. Intrigued by the son of four-time leading sire Leamington after watching the horse race at Pimlico, Hancock was initially unable to meet the asking price for him, but the desire to have him was so strong that he spent considerable time tracing his whereabouts. He eventually got him in exchange for the stallion's half-brother, Scathelock, which Hancock had received as a gift. The trade paid off, as his grandson raced to victory in the Preakness wearing Ellerslie's colors. Arthur Hancock Sr. sought the stallion Celt, which he lost in an attempted long-term lease arrangement; he eventually purchased him for $25,000, outbidding many well-to-do horsemen. Celt went on to lead the sire list in 1921 and sired thirty stakes winners.

Bull first had the British Bloodstock Agency make a bid of $50,000 for Nasrullah, right after his retirement from racing, but it came too late. Irish politician Joseph McGrath had purchased him just the week before. In 1949, Hancock tried once again to acquire him. Heading a syndicate with Harry F.

Guggenheim, Mr. Woodward and Canadian Eddie (E.P.) Taylor,[64] they were to make an offer of £100,000.

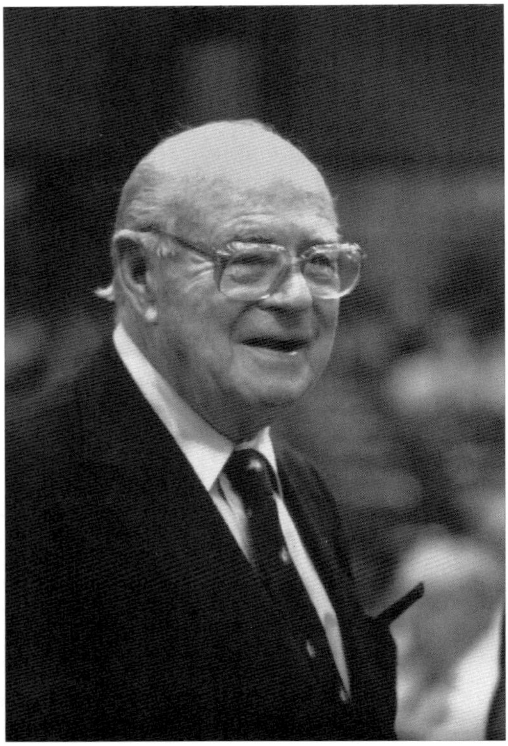

Canadian businessman and horseman E.P. Taylor, a member of the syndicate that purchased Nasrullah. *Courtesy of the Canadian Horse Racing Hall of Fame.*

Edward Plunket (E.P.) Taylor is remembered as one of Canada's most famous businessmen and beloved horsemen. Twice crowned outstanding breeder in both Canada and the United States, "Eddie," as he was known to his friends, started in the brewing business, having inherited Brading Brewery from his grandfather. He was a volunteer for the Canadian government in World War II and was appointed by Winston Churchill to the British Supply Council in North America. E.P. got his start in racing in the 1930s with his Cosgrave Stable, and his Windfields Farm would later be responsible for breeding Northern Dancer, the greatest sire of the twentieth century and son of Nearctic, another Nearco stallion. One of his first forays into breeding came with his involvement with the Americans who set out to purchase Nasrullah.

At the time Bull made his attempts to purchase the stallion, Nasrullah was standing for a stud fee of $1,120. Complications arose, however, when the English currency was devalued, and McGrath would not let Nasrullah go. According to Hancock:

> *We dealt entirely in pounds. Eddie was going to pay for the horse and we were going to re-pay him, because he had pounds and wanted to get some dollars in exchange. Everything seemed in good order; but Eddie called me one night and said, "We're not going to get that horse for 100,000 pounds." I asked him why, and he said, "The day after tomorrow they are*

going to devalue the pound. That man isn't going to let him go." That's just what happened.[65]

In 1950, McGrath agreed to a deal of $320,000, but there was yet another setback. The "earnest money" to secure the deal did not arrive in time as a result of a bank holiday, so McGrath thought the deal was off. In an interview, Bull claimed that McGrath had had a bigger offer from California. It was rumored that the offer came from film industry lawyer and racehorse owner Neil McCarthy,[66] but for unknown reasons McGrath was looking for a way out of the deal.

Perhaps drawing inspiration from his grandfather's quest for Eolus, after previous failed attempts, Bull was more determined than ever to lay claim to the stallion, and he set off for New York (accompanied by his wife and her father, Seth Walker, an attorney), where he met with McGrath and was able to work out a deal of $340,000[67] with McGrath retaining an annual breeding right.

It had taken Bull a full seven years to get the stallion in his possession, a year longer than it took his grandfather to acquire the colt Eolus, but it was all well worth it in the end. As turf writer Edward L. Bowen proclaimed, "Getting a new stallion is like Christmas to horse breeders, but getting a grand animal like Nasrullah was Christmas times ten."[68]

Nasrullah would go on to establish a breeding dynasty that not only revived Claiborne and returned the farm to its former place of prominence atop the Thoroughbred industry but would also see Kentucky continue its rise to glory as the state of champions, which, if his descendants have anything to say about it, will likely continue long into future generations of the breed.

Chapter 6

CLAIBORNE DAYS

Nasrullah was set to arrive in the Unites States on Independence Day 1950. He set his hooves on Kentucky bluegrass after a lengthy journey by ship from Ireland and then by train from New York. An article in the *Canberra Times* on Wednesday, July 5, 1950, included a write-up of the stallion's arrival in America:

> *New York, Tues. The famous Irish stallion, Nasrullah, valued at £167,411 arrived here yesterday after a 12-day Atlantic crossing. He is headed for the stud at Paris, Kentucky. Nasrullah sired Noor, who has beaten the American champion Citation four times in as many meetings.*[69]

Footage released in 2010[70] by the Claiborne estate shows the stallion first stepping off the train onto American soil and then being led out of the van right into his new paddock, where he frolicked and romped around on the day of his arrival.

Nasrullah's arrival was well remembered by Kentucky breeders. As Bob Courtney, owner of Crestfield Farm in Lexington, recalled:

> *When Nasrullah first was imported, the talk on the street was focused on wondering why the English would let the horse go, but once he got started on the racetrack, he set the world afire. On the farm, Nasrullah was surely a handful to work with. He'd get up on his hindlegs and take out lightbulbs or anything else, but man, he was some kind of sire.*[71]

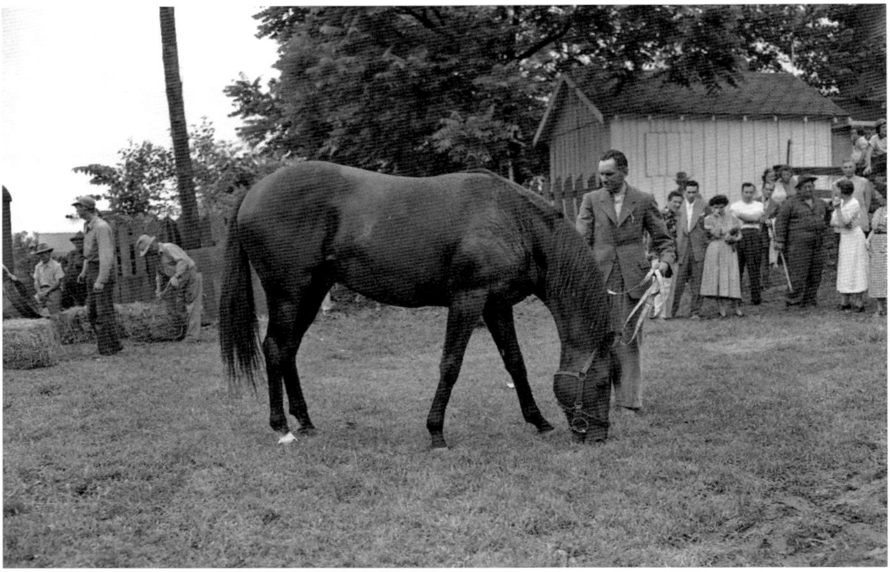

Nasrullah arrives at Claiborne, July 4, 1950. *Courtesy of Keeneland/Meadors.*

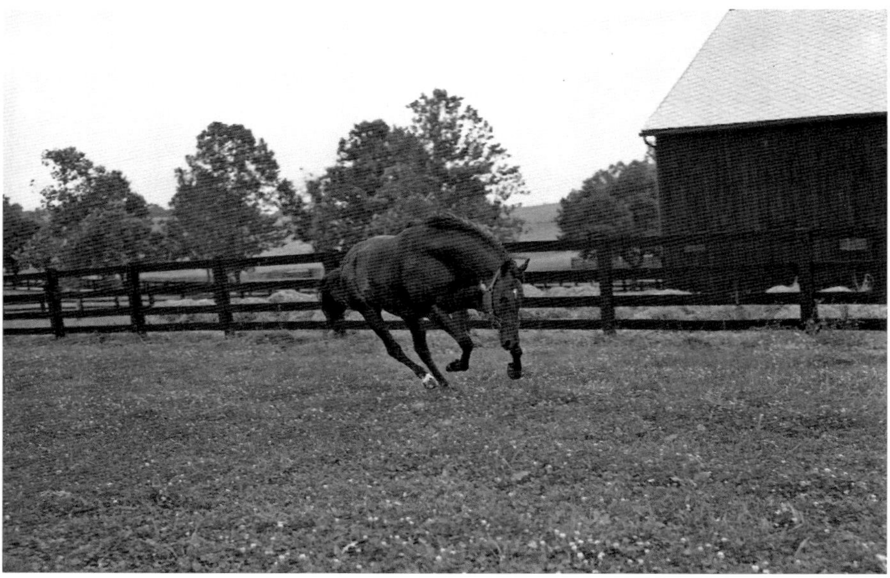

Nasrullah romps around his paddock on the day of his arrival at Claiborne. *Courtesy of Keeneland/Meadors.*

McGrath's stud groom, Bill Milton, stayed on at Claiborne until Nasrullah got settled in; the horse was then placed in the care of stallion groom Lawrence Robinson. Mrs. Hancock observed Milton's method of getting Nasrullah in hand: approaching the stallion with a stick, which he was expected to latch onto before being led away. She recalls that Robinson was adamant that this was simply not his way of handling horses. Bull assured him that as he was responsible for the stallion's care, he could refrain from this approach, saying, "Lawrence, you're in charge. Just do it your way."[72]

Robinson and Nasrullah eventually forged a connection; the veteran stud groom described it as a "love/hate relationship."[73] Farm workers certainly had their share of stories to tell of the stallion's antics. Colonel Floyd Sager, Claiborne's veterinarian, reportedly never did give the horse his tetanus shot because of his unruly behavior. After hearing Sager's complaints about Nasrullah's temper, an exasperated Bull Hancock said, "Well, Colonel, I guess we'll just have to put up with him, won't we?"[74] Bull himself reportedly threatened the stallion with a broom in the breeding shed after he got out of the control of his handlers.[75]

Nasrullah stood nine seasons at Claiborne Farm. He shared the stallion barn with such champions as Kentucky Derby–winner Dark Star; imports Ambiorix, Blenheim II and Princequillo; the latter's son Hill Prince; and successful sire Turn-to. He spent his mornings grazing and frolicking in his own paddock, and the rest of his time was spent sleeping and being paraded around for visitors.

"He's actually a pretty spoiled horse," Bull Hancock was quoted as saying.

> *If visitors want to look at another stallion first, he'll kick up a hell of a fuss in his stall as if to say, "There's no point in looking at those bums when you can look at me." In a way perhaps he's right. I think he's the grandest looking horse I ever saw.*[76]

Nasrullah quickly became highly regarded as a stallion in the United States; his influence on the modern American Thoroughbred is undeniable. Kent Hollingsworth, writing in *The Kentucky Thoroughbred*, is of the opinion that "Nasrullah was the most influential, the most important stallion imported to America since the first English Derby winner, Diomed, arrived in Virginia in 1800."[77]

Before any of his American offspring were born, let alone set foot on a racetrack, Nasrullah was making a name for himself as a notable sire through the successes of his sons and daughters overseas. In 1950, the year he was

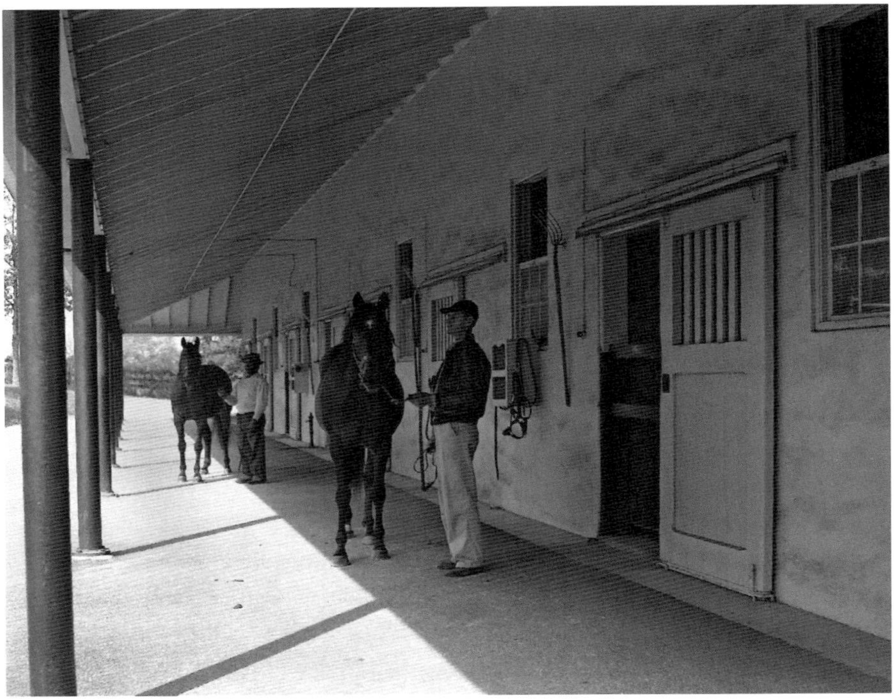

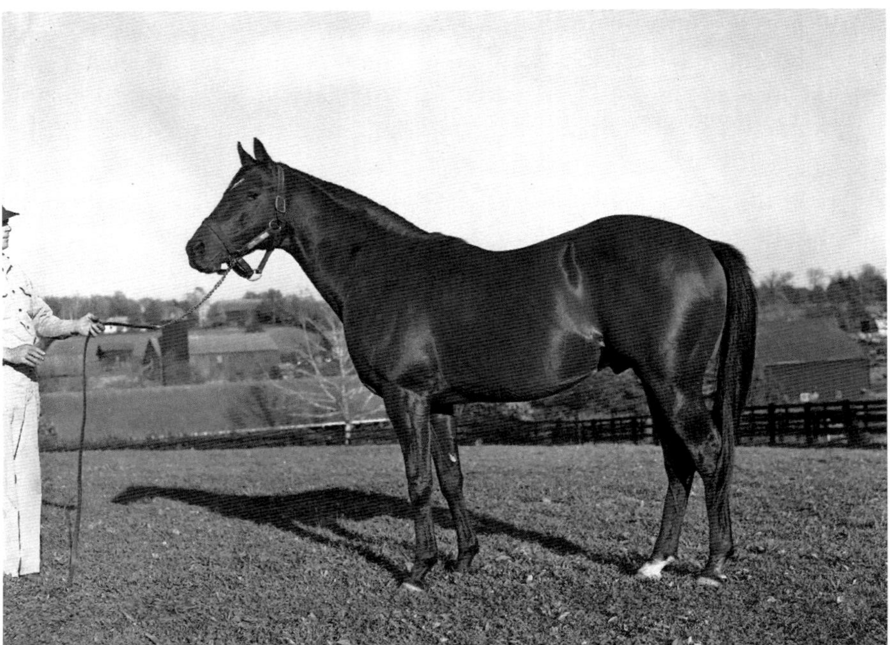

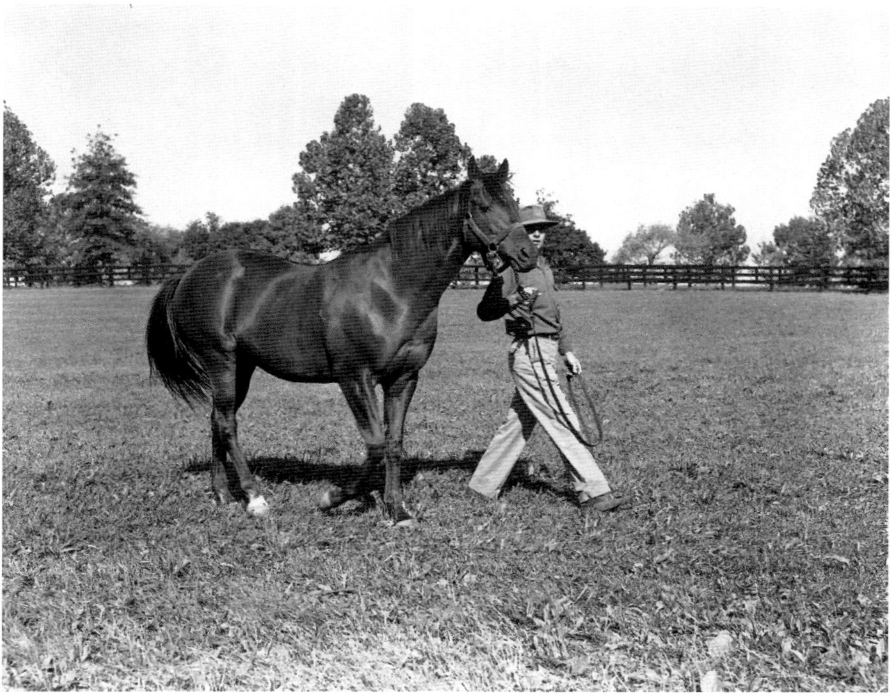

Above: Nasrullah, 1951. *Courtesy of Keeneland/Meadors.*

Opposite, top: Nasrullah (front) and Princequillo (back) at Claiborne, 1950. *Courtesy of Keeneland/Meadors.*

Opposite, bottom: Nasrullah at Claiborne, 1950. *Courtesy of Keeneland/Meadors.*

imported, Belle of All was voted the best English juvenile filly after capturing the Cheveley Park Stakes. The next year, she won the 1,000 Guineas. His son Noor caught everyone's attention in his stellar 1950 campaign in America, highlighted by his four consecutive triumphs over Citation. Back in England, Nearula was the high-weighted juvenile colt that year; he followed this up with a win in the 2,000 Guineas in his three-year-old season. In 1951, his first stud season in Kentucky, Nasrullah led the sire list in England.

Nasrullah's first American-conceived foals arrived in 1952, when he was twelve. That year, he was in third place on the sire list in England. His first American crop consisted of thirty-two foals, and they were considered hot commodities. Only two sold at auction as yearlings, commanding $17,500 and $16,500 at the Keeneland summer sale. Eight went on to become stakes winners (25 percent): Nashua, Delta, Flying Fury, Blue Ruler, Insouciant, Lea Lane, Jean's Joe and Independence. Courtesy and Natasha were also stakes placed. Courtesy later became the dam of three stakes winners—Knightly Manner, Dignitas and Respected—all by the top Claiborne stallion Round Table, and she is the third dam of champions Swale and Forty Niner. Natasha was out of the Belair champion Vagrancy, and she is best known as the dam of stakes-winner Natashka, a Kentucky Broodmare of the Year.

Nashua was his sire's leading money-earner at ages two, three and four and led the nation in earnings as a three-year-old. Racing for Maryland's Belair Stud, he captured two legs of the Triple Crown in winning the Preakness and Belmont and later defeated Kentucky Derby–winner Swaps in a match race. Nashua was the only champion from this crop and the first of eight American-born champions by Nasrullah, but the rest were successful racers and producers in their own right.

Claiborne-bred Delta raced through the ages of two to four, defeating half-sister Lea Lane on two occasions as a two-year-old and winning a total of twelve times, including four stakes races as a four-year-old. She became a top broodmare for Claiborne, producing five stakes winners, and was voted 1968 Kentucky Broodmare of the Year.

Flying Fury was a stakes winner for Harry F. Guggenheim's Cain Hoy Stable at ages two, three and four and at stud sired seven stakes winners. Blue Ruler, bred by Claiborne, raced in California for Murcain Stable and won several stakes as a two-year-old. Insouciant also won the Ashland and Coronet Stakes and the Modesty Handicap for owner Dr. Eslie Asbury.

Charlton Clay's Lea Lane won three stakes at two; at three, she ran second in the Kentucky Oaks and third in the Ashland. As a broodmare, she produced stakes-winner Rose Bower and is the third dam of top sire Miswaki. Hungarian-born John Hertz, founder of the Yellow Cab Company, raced the colt Jean's Joe, which was stakes-placed three times as a two-year-old and won the San Felipe Handicap in his three-year-old season. He was also runner-up in the Santa Anita Derby, Blue Grass Stakes and Derby Trial.

Independence was bred by the Phippses' Wheatley Stable and was a full brother to Bold Ruler. He became a top steeplechaser for Mrs. M.G. Walsh, racing for eight years, breaking track records at Belmont Park and Saratoga and earning $132,088 in seventy-four starts. Among his wins was the 1961 Grand National Steeplechase Handicap at Aqueduct. He sired Canadian equestrian team horse Catbird and Puissance-winner Indigo.

Nasrullah was second on the English sire list in 1953. Star foals from his second American crop included co-champion two-year-old filly Nasrina, winner of the Frizette and Gardenia Stakes races for owner Howell E. Jackson, and Glamour, which established herself as a top broodmare, producing the stallion Poker.

Nantallah is an interesting case. A mediocre runner, he returned to Claiborne, where he proved himself at stud, much as his own sire did. Matched on numerous occasions with the mare Rough Shod II, the two became quite the dynamic duo, producing four outstanding offspring. Daughter Thong became dam of Special, which in turn produced Nureyev, champion three-year-old colt in France and sire of Breeder's Cup Turf–winner Theatrical, as well as Fairy Bridge, which is the dam of Sadler's Wells, champion miler in France and a leading sire in the United Kingdom. Nantallah and Rough Shod also produced Moccasin (champion two-year-old Filly and Co-Horse of the Year and dam of seven stakes winners); Ridan, unbeaten at two, winner of the Florida Derby and Blue Grass Stakes at three and archrival of Nasrullah's son Jaipur; and the sprinter Lt. Stevens.

In 1953, Claiborne chose not to send any of its yearlings to auction, opting instead to sell a select few privately and to keep the rest for racing, part of an "intensive rebuilding program."[78] From this crop, four of Claiborne's Nasrullah yearlings were sold, including the aforementioned Blue Ruler. The fillies Delta and Courtesy were among those that Claiborne wisely retained; these two would go on to contribute substantially to the farm's breeding program.

In 1954, a Nasrullah yearling Nalur, out of the mare Lurline B, sold for a record $86,000 at the spring sales at Keeneland. This price surpassed the $75,000 that was put down for the colt New Broom in 1928. Unfortunately, Nalur did not live up to expectations; his career earnings totaled a meager $6,575. He proved to be an outlier in the Nasrullah dynasty.

Nasrullah went on to be the leader by average price at the yearling auctions in 1954 and 1955. With the market booming at the time, total sales in both years were records, and the average price of 1955 yearlings was the second highest on record. From 1953 through 1958, thirty-two Nasrullah yearlings were sold at public auction for a total of $1,164,100, with an average of $36,378.[79]

Nasrullah's first American foals began racing in 1954. He also led the list of sires of two-year-old money-winners that year. The *Thoroughbred Record* reported that the $625,692 earned by Nasrullah's first American-bred crop of two-year-olds in 1954 was an all-time record.[80] He was also second to Hyperion on the sire list in England. This was the year that Never Say Die raced to victory in the Epsom Derby, becoming the first American-bred to win the race since Iroquois in 1881 and the first derby winner for eighteen-year-old jockey Lester Piggott. Later that year, Never Say Die won the St. Leger.

In an article written for *Sports Illustrated*, Whitney Tower proclaimed of Nasrullah, "That he is sire of the year is undisputed. He may well be the great sire of this decade."[81] He was certainly the most sought-after stallion through the remainder of the 1950s.[82]

Bold Ruler, arguably Nasrullah's most famous and influential son; Leallah, co-champion two-year-old filly in 1956; and eventual top sire Red God were notable foals from the stallion's third crop. Reportedly ill tempered, Red God was out of the mare Spring Run and was foaled at Claiborne. Owner Harry F. Guggenheim sent him to race in England, where he won the Richmond and was second in the Champagne Stakes as a two-year-old. At four, the chestnut returned to the United States to take the Roseben Handicap. Standing at Loughton Stud in Ireland, Red God sired ten stakes winners and established one of the preeminent branches of the Nasrullah line through his son Blushing Groom.

Other progeny of 1954 included Cain Hoy Stable's top handicapper One-Eyed King, repeat winner of the Donn Handicap, and the successful California stud Nashville, which twice defeated Bold Ruler.

In 1955, Nasrullah became the first stallion to lead both the English and American sire lists. He was also the leading juvenile sire in North America for the second time, aided by Nasrina's stellar two-year-old campaign.

That year, three of his sons—Nashua, Jean's Joe and Flying Fury—started in the Kentucky Derby. Swaps took the win over Nashua, with Jean's Joe and Flying Fury finishing fifth and sixth, respectively. Also that spring, three of Nasrullah's daughters put on a show in the Ashland Stakes for three-year-old fillies at Keeneland. Dr. Eslie Asbury's homebred Insouciant finished first, followed by Claiborne's Courtesy, with Lea Lane back in third.

Foals of 1955 included Eclipse champions Bald Eagle and Nadir. A horse from this crop that William Woodward might have regretted letting go is Bug Brush. Out of Belair Stud's stakes-winning mare Bonnie Beryl, Bug Brush won the Kentucky Oaks at three and was a top handicapper at four, taking the San Antonio Handicap over colts Hillsdale, Terrang and half-brother Fleet Nasrullah, setting a world record of 1:46 $^2/_8$ for the mile-and-an-eighth race. Owner C.V. Whitney called her his greatest distaffer ever.

Fleet Nasrullah was also a winner—both on the track and in the breeding shed. The story of the colt's racing success is an amusing one and demonstrates the uncertainty of the racing business. The story goes that owner/breeder John Hertz didn't believe the horse could go longer than six furlongs and sold him for $165,000 to Ellwood Johnston, who had faith in the colt's capabilities beyond sprint distances and boldly proclaimed that he would prove it. Hertz reportedly smiled and said, "You're entitled to your opinion." Fleet Nasrullah went on to set a new track record for nine furlongs on the turf and win the mile-and-one-sixteenth Californian. Hertz tried to buy him back for $750,000, but Johnston turned him down.[83] He retired the horse to his Old English Rancho in California, where he sired twenty-six stakes winners between 1962 and 1965. He was then syndicated for $1.2 million and sent to Spendthrift Farm in Lexington, where he sired another fifteen stakes winners and stood until his death in 1979.

Other stakes winners from the 1955 crop included Victory Morn, Sir Ruler and Nasco, a full brother to Bold Ruler. Steeplechaser Nala and Galarullah, a $65,000 yearling purchase and runner-up in the Kentucky Oaks, were also notable also-rans.

Total earnings for Nasrullah's progeny in 1955 was $1,433,660, a record that had been surpassed only once before. Nashua took home a fair share of these earnings; he was crowned Horse of the Year as well as champion three-year-old colt. A 1956 stud ad from Claiborne Farm proclaimed that Nasrullah had fifty-six stakes winners in Europe and America to date at the time.

In 1956, Nasrullah led the American sire list for the second consecutive year. He was also the Leading Juvenile Sire in North America that year for the second time; this achievement owes much to the success of co-champion two-year-old filly Leallah and stakes winners Bold Ruler and Red God.

On-and-On was surely the gem of the 1956 Nasrullahs. Out of the Bull Lea mare Two Lea, he had very successful racing campaigns at three and four, running under the "renowned devil red and blue silks" of Calumet Farm—in its heyday one of the premier Thoroughbred establishments in Kentucky. Half-brother to Kentucky Derby–winner Tim Tam, On-and-On sired derby- and Preakness-winner Forward Pass. He is also the broodmare sire of Alydar and sired Grand Prix jumpers Coach Stop and Everest Philco.

Following up on the success of his half-brother and stable mate On-and-On, Pied D'Or emerged as a stakes winner from Nasrullah's next crop and is joined by Virginia-bred Grand Prix jumper Nanticoke along with Kentucky-bred stakes-winners Nascania and Lurullah as the stars of 1957. That year, at age seventeen, a single share in the syndicated stallion, a $^{1}/_{34}$ interest, cost $65,000. Nasrullah's market value at that time was estimated at $2,210,000.[84]

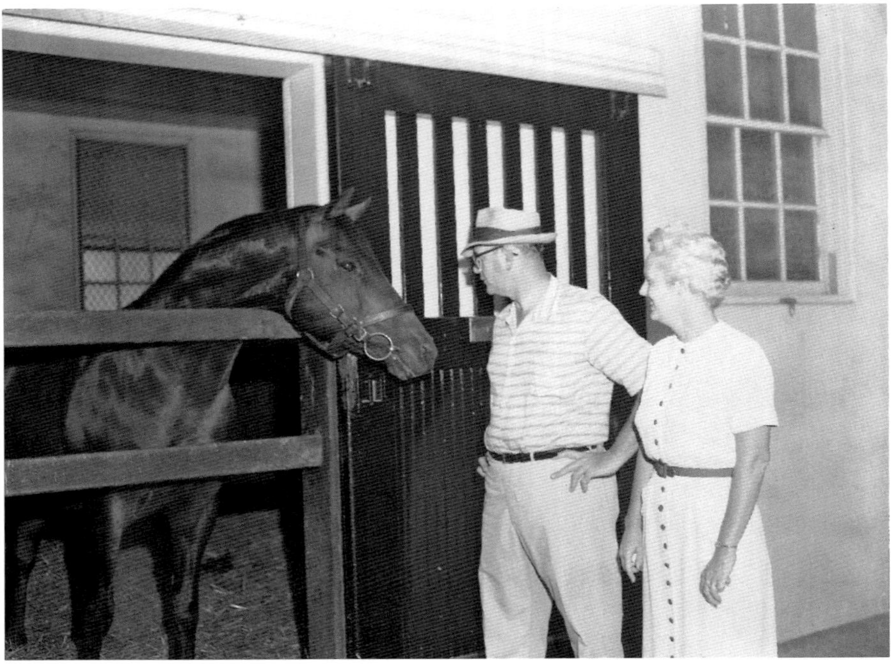

Nasrullah with Mr. and Mrs. Hancock, 1957. *Courtesy of Keeneland/Meadors.*

As William Nack, author of *Secretariat: The Making of a Champion*, explains:

> *He* [Nasrullah] *represented a milestone in financial investment in racing blooded horses, the beginning of big profits from what were only the most modest investments. An original $10,000 investment in a share of Nasrullah ultimately became worth about $700,000.*[85]

Two sons of Nasrullah took home three Eclipse Awards in 1957. Bold Ruler claimed Horse of the Year and champion three-year-old colt honors, and Nadir was crowned champion two-year-old colt.

A Claiborne colt by the name of Ambassador expanded Nasrullah's influence beyond the racing oval. Unsuccessful on the flat, the gray excelled over jumps and became a stakes-winning steeplechaser for the farm. While doing an interview for *Sports Illustrated*, Bull Hancock received a phone call in which he learned that Ambassador had won a 'chase at Belmont. Laughing, he recounted the colt's initial failings until they discovered his knack for jumping:

> *Once I talked seven guys I play golf with in Florida into betting $100 each on him, and he wasn't even in the hunt. Now suddenly, after he can't catch anything in front of him on the flat, he sees something in front of him on a damn jump course, and he not only jumps it, but they can't catch him.*[86]

Nasrullah's seventh American crop of foals, which arrived in 1958, included the Wheatley Stables' Misty Day, eventual top broodmare, and her stakes-winning stable mate Royal Record. Bold Ruler returned to the stage to lay claim to champion sprinter honors. That year, Nasrullah's North American get had earned $6,519,175 to date.[87]

Jaipur was the star of Nasrullah's 1959 crop. Winner of multiple stakes races, including the Belmont and Travers, he was crowned champion three-year-old colt of 1962. Eventual top jump sire Gayrullah was also a foal of 1959. His son Brother Sam, ridden by Canadian show-jumping rider Ian Millar, won an individual bronze medal and took team silver at the 1979 Pan Am Games. Claiborne stakes-winner Nas-Mahal would go on to produce champion three-year-old filly Turkish Trousers and Hollywood Gold Cup–winner Tell. Intercepted would establish himself as a winning handicapper for Pin Oak Stud.

At year's end, Nasrullah led the American sire list for the third time. But the best, and the worst, was yet to come.

Chapter 7

A LEGEND DIES, A LEGACY IS BORN

At about 9:00 a.m. on May 26, 1959, several hours after being turned out in his paddock at Claiborne, nineteen-year-old Nasrullah began behaving strangely. Grooms Snow Fields and Lawrence Robinson, who were in the breeding shed with Bull Hancock, were alerted to Nasrullah's distress by the frantic nickering of Bold Ruler, which was turned out in the paddock alongside his sire. The men rushed outside to find the stallion sweating and staggering. Colonel Floyd Sager, the farm's resident veterinarian, was called, but not long after his arrival, Nasrullah died, at approximately 10:20 a.m.

An autopsy performed at the University of Kentucky later revealed that the cause of death was a ruptured heart. The left ventricle of the horse's heart burst, causing blood to pour into and fill his thoracic cavity, leading to suffocation. Robinson described himself as being "surprisingly choked up" by the stallion's death,[88] and similar feelings were expressed by those who had known the horse.

Nasrullah became the fifth horse to be buried in the cemetery at Claiborne, his headstone placed beside those of Sir Gallahad III, Johnstown, Gallant Fox and Blenheim II. Son Nadir was retired to his sire's stall.

At the time of his death, Nasrullah had covered thirty-six of the thirty-seven mares booked to him that season. At the end of 1959, he was leading sire for the third time, with his runners earning $1,434,543 on American racetracks. He was considered to be one of the most valuable Thoroughbreds in the United States at the time.[89] Money earned that year by horses bred and raised at Claiborne totaled more than $4 million—a new record.[90]

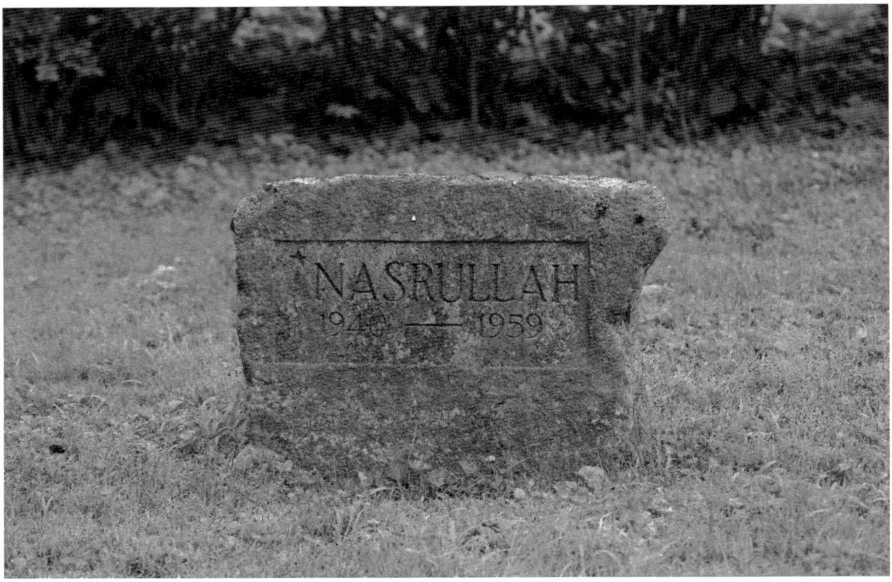

Nasrullah's final resting place at Claiborne Farm. *Courtesy of Jessica Morgan.*

Nasrullah's final crop was born in 1960, in the spring following the stallion's death. He led the sire list for the fourth time that year, helped in part by Bald Eagle's being crowned champion in the older male division.

From this crop came champion Never Bend and a number of stakes winners, including Nasram II. Bred in Kentucky and racing for Mrs. Howell E. Jackson, he captured the Prix du Lys in France as a three-year-old and scored one of the biggest upsets in British racing history when he defeated Epsom Derby–winner Santa Clause in the King George VI and Queen Elizabeth Stakes as a four-year-old. Nasram II stood at stud in both the United States and Germany; his son Naskra, fourth in the Kentucky Derby, became sire of champion sprinter Star de Naskra and grandsire of stakes-winner Carr de Naskra.

Also from Nasrullah's final crop came Greentree Stables' Outing Class, which won the Hopeful Stakes as well as a number of handicaps but is better known as the grandsire of Canadian broodmare of the year No Class, dam of Canadian champions Sky Classic, Regal Classic, Grey Classic and Classy 'n Smart (grand-dam of 1991 Canadian Horse of the Year Dance Smartly), as well as sire Smart Strike. There was also the Kentucky-bred Poster Girl, which became the dam of Talking Picture, 1973 U.S. champion two-year-old filly.

In 1962, Nasrullah led the North American Juvenile Sire list for the third and final time, based largely on Never Bend's recognition as champion two-year-old colt. He also led the American sire list again (three times posthumously), for a total of five times. Only two other horses matched this achievement in the twentieth century—Star Shoot and Bull Lea—and Nasrullah's own son Bold Ruler would surpass it, going on to lead the sire list an astonishing eight times.

In total, Nasrullah sired 245 American starters that earned $6,519,175—192 were winners, 93 stakes winners, 9 champions and 3 Hall of Famers. All told, Nasrullah sired 98 stakes winners (23 percent) from 422 named foals; 10 percent is generally considered to be exceptional for a stallion. His offspring won 1,088 races and $12,810,228.

Impact on the Classics

Nasrullah's offspring had early success with the European classics. He is one of only five horses to sire winners of all five of the British classics—the 2,000 Guineas, 1,000 Guineas, Epsom Derby, Epsom Oaks and St. Leger Stakes—and in fact he got six winners. Never Say Die won the 1954 derby and also captured the St. Leger. Grandsons Larkspur, Mill Reef and Roberto were also victorious in the derby. Son Nearula won the 2,000 Guineas, while daughter Musidora and granddaughter Never Too Late II (Never Say Die) got both the 1,000 Guineas and the Oaks. Belle of All also won the 1,000 Guineas.

While Nasrullah sired the winners of numerous classic races, both internationally and in the United States, he failed to produce a winner of the Kentucky Derby, despite having a significant number of starters in the race. Sons Nashua and Never Bend came closest to victory, with Nashua finishing second to Swaps in the eighty-first running in 1955 and Never Bend placing in the eighty-ninth running in 1963, behind Chateaugay.

In 1955, two other sons of Nasrullah started in the Run for the Roses: Jean's Joe and Flying Fury, which finished fifth and sixth, respectively. Bold Ruler ran fourth in the 1957 derby, before capturing the other two legs of the Triple Crown. In the derby of 1958, grandson Noureddin (Noor) placed third, and son Martin's Rullah finished fifth. Foreign-bred Die Hard (Never Say Die) ran sixteenth out of seventeen starters in 1959, while Lurullah ran twelfth in 1960. In 1962, grandson Ridan finished third. In 1965, three grandsons ran: Swift Ruler, Bold Lad and Narushua, finishing seventh, tenth and eleventh, respectively.

Bull Hancock's greatest wish was to win the Kentucky Derby; he once said, "I'd rather win a Kentucky Derby than all the rest of the races in the United States put together."[91] With his stables full of Nasrullah's colts and fillies, his chances of capturing the elusive title looked promising. Though he thought he had a chance in 1958, his hopes were dashed when Nadir failed to win a prep race and was withdrawn from the race. Sadly, Bull's dream died with him; it would be 1984 before Claiborne got its first derby winner with Swale, a direct male-line descendant of Nasrullah.

The first descendant of Nasrullah to actually win the derby was Forward Pass, a colt by On-and-On, but he was only placed first in the 1968 running after the winner, Dancer's Image, whose maternal grandsire was Noor, was disqualified. Also in the derby field that year (and their placings) were great-grandsons T.V. Commercial (fourth), Don B. (seventh), Te Vega (tenth) and Verbatim (thirteenth) and grandson Iron Ruler (twelfth). In 1969, grandsons Dike and Fleet Allied both ran, finishing third and seventh. Nasrullah's influence on the race over the last forty-plus decades has been immense. Beginning in 1970 with Dust Commander, a descendant of Nasrullah has won thirty-seven of the last forty-two runnings of the Kentucky Derby.[92]

Bold Ruler had more success with the derby than his sire, producing a winner in Secretariat and becoming the grandsire of five other winners. Nasrullah was more successful in the other two legs of the Triple Crown, siring two winners of the Preakness, Nashua and Bold Ruler, and two winners of the Belmont, Nashua and Jaipur.

But the progeny of Nasrullah were not only celebrated in the winner's circle. He established a breeding legacy that remains largely unprecedented to this day. He or one of his descendants topped the national sire ranks a whopping eighteen times from 1955 to 1984. In 2007, A.P. Indy became the seventh male-line descendant of Nasrullah to head the North American sire list.[93] Nasrullah was also broodmare sire of 159 stakes winners, including champions Creme dela Creme, Dike, Drumtop, Fiddle Isle, Hornbeam, Lacquer, Natashka, Sovereign II, Talking Picture, Tell and Turkish Trousers.

Notable male lines descending from Nasrullah are those of Bold Ruler, Nashua, Never Bend, Grey Sovereign and Red God. The influence of each stallion is profiled below.

Bold Ruler

Bold Ruler is generally regarded as the most memorable and influential of Nasrullah's sons, as he made a splash both on the track and at stud. A winner at varying distances, he passed on both speed and stamina to his own progeny. Standing at Claiborne, he led the general sire list a total of eight times (1963–69, 1973). From just 356 foals, he sired 82 stakes winners (23 percent); 10 of them were crowned champions in the United States.

Secretariat is undoubtedly the most famous of Bold Ruler's offspring. Winner of the 1973 Triple Crown, five Eclipse Awards (he was crowned Horse of the Year twice, both as a two- and three-year-old) and a Hall of Fame inductee, "Big Red" was successful as a sire as well. He was the 1978 leading sire of two-year-olds and Leading Broodmare Sire in 1992. His major stakes winners include Horse of the Year Lady's Secret; Preakness- and Belmont Stakes–winner Risen Star; Tinner's Way, which won consecutive runnings of the Pacific Classic; and Travers Stakes–winner General Assembly. He is the damsire (maternal grandsire) of top stallions A.P. Indy, Storm Cat and Gone West. In Secretariat and his resulting offspring, Bold Ruler dispelled the myth that he was not a producer of true stayers.

Bold Ruler led the list of sires of two-year-olds a total of six times. Champion two-year-olds sired by Bold Ruler are the fillies Queen Empress (1964) and Queen of the Stage (1967); two colts that share the name of Bold Lad (1964, 1966);[94] and colts Successor (1966), Vitriolic (1967) and Secretariat (1972).

Other champions sired by Bold Ruler are the ill-fated filly Lamb Chop, which broke down in the 1964 Strub Stakes and was crowned champion three-year-old filly posthumously; Bold Bidder (1966 champion older male and sire of Kentucky Derby–winners Cannonade and Spectacular Bid—the latter also a four-time Eclipse Award winner); Gamely (three-time Eclipse award winner); and Wajima (champion three-year-old colt in 1975). Seven of Bold Ruler's male-line descendants captured the derby in the 1970s.

Bold Ruler is also grandsire to innumerable champions and stakes winners. Son What a Pleasure, winner of the Hopeful Stakes at two, was leading sire in 1975 and 1976. Among What a Pleasure's top runners are champion two-year-old colts Foolish Pleasure (Kentucky Derby), Honest Pleasure and the stakes-winning mare Gene's Lady. Irish Castle, also a winner of the Hopeful Stakes, sired Bold Forbes, 1976 Kentucky Derby and Belmont Stakes winner. Bold and Brave sired the stakes-winning filly and Hall of Fame inductee Bold 'n Determined. Son Reviewer, which

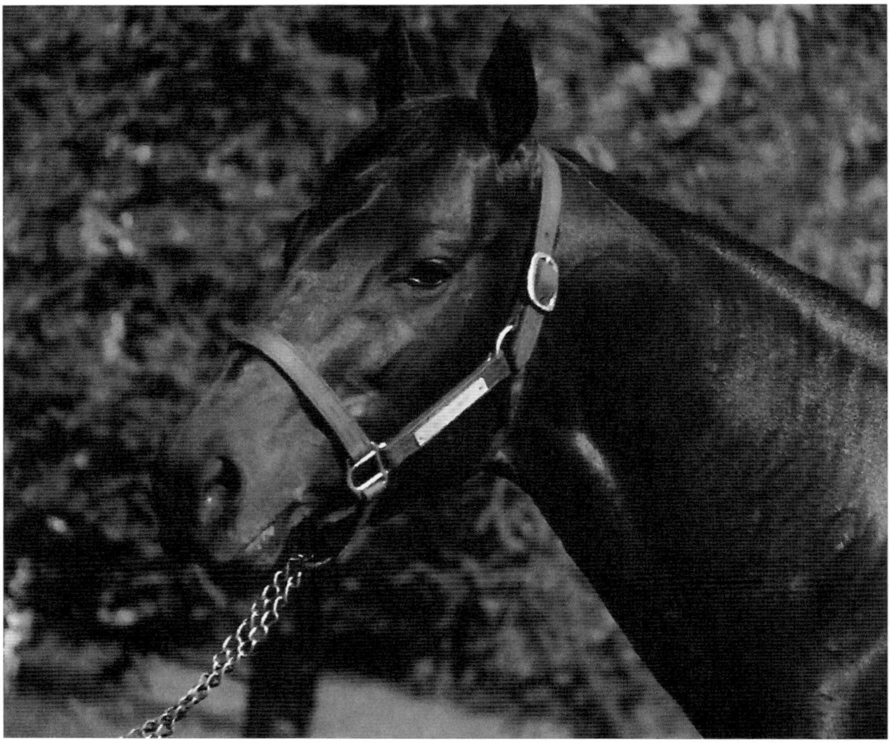

Leading Canadian sire Bold Ruckus, a direct male-line descendant of Nasrullah through Bold Ruler. *Courtesy of the Canadian Horse Racing Hall of Fame.*

suffered three cannon bone fractures while racing, sired champion filly Ruffian, which inherited her sire's weak bones and broke down in a match race against another Bold Ruler grandchild, Foolish Pleasure, as well as champion three-year-old filly Revidere.

Boldnesian, while not a great success on the track, sired stakes-winner Bold Reasoning, which in turn sired Triple Crown–winner Seattle Slew, and Bold Ruckus, a Canadian Horse Racing Hall of Fame inductee that dominated the sire lists north of the border in the 1990s, siring two-time Sovereign Award–winners Bold Ruritana, King Corrie and King Ruckus and two-time leading sire Bold Executive, sire of three Canadian champions (Sand Cove, Blonde Executive and Simply Lovely).

Bold Ruler also established himself as a sire of leading juvenile sires. Beginning in 1970 with Cornish Prince, a son of Bold Ruler was among the top-ten sires of juvenile earners for fourteen consecutive years; on five occasions during that time, a son led the list (What a Pleasure in 1974 and 1975, Raja Baba in 1976 and 1980 and Secretariat in 1978). Two of Bold Ruler's sons led the general sire list (What a Pleasure in 1975–76 and Raja Baba in 1980).

The Bold Ruler sire line continues to dominate in the twenty-first century, primarily through the branches of Seattle Slew (led by A.P. Indy) and Boldnesian (led by Bold Ruckus). Sons Raja Baba (Yes It's True and Notebook); Chieftain (Fit to Fight); What a Pleasure (Mecke); Secretariat (Tour d'Or); and Dewan (Glitterman) have also established influential lines.

Nashua

At stud, Nashua sired a total of seventy-seven stakes winners (12 percent). He sired three winners of the Coaching Club American Oaks in Bramalea, Marshua and Shuvee. The latter won the New York Fillies Triple Tiara (formerly known as the Filly Triple Crown, composed at the time of the Acorn Stakes, the Mother Goose Stakes and the aforementioned Oaks), was Champion Older Mare in 1970 and 1971 and was a Hall of Fame inductee in 1975. Shuvee remains the only filly to have won the Jockey Club Gold Cup, run at two miles at the time, which she captured twice, following in the footsteps of her sire. Other noteworthy offspring of Nashua include Swaps Stakes–winner Noble Nashua and Diplomat Way, winner of the 1966 Arlington-Washington Futurity, the country's richest race at the time, and grandsire of Horse of the Year Skip Away.

Nashua never led the sire list but was second in 1964 and ranked in the top five on two occasions. Nashua is best known, however, as a broodmare sire. His daughters produced 123 stakes winners. Daughters Gold Digger and Bramalea have had the most impact in establishing lasting sire lines from this branch. Gold Digger, out of the mare Sequence, was a stakes winner for Spendthrift Farm. She became the dam of Mr. Prospector, the all-time leading sire of stakes winners, twice the leader of the U.S. sire list and leader of the broodmare sire list nine times. Mr. Prospector established notable sire lines of his own in sons Fappiano, Forty-Niner, Gone West, Gulch and Seeking the Gold. Bramalea, out of the mare Rarelea, produced Epsom Derby–winner Roberto, a three-time champion, sire of six champions and a leading sire in Europe and North America.

NEVER BEND

Never Bend sired a total of sixty-one stakes winners (17 percent). The most famous of his offspring is Mill Reef, European Horse of the Year, champion three-year-old in England and champion older horse in 1972. Among his wins were the Epsom Derby and the Prix de l'Arc de Triomphe. Mill Reef sired Epsom Derby–winners Shirley Heights and Reference Point and Glint of Gold, stakes winner in multiple countries.

Mill Reef has left a legacy in the Breeder's Cup Turf that is nothing short of phenomenal. Son Lashkari won the inaugural running of the Breeder's Cup Turf in 1984; grandsons Kotashaan, Daylami and Kalanisi captured the 1993, 1999 and 2000 runnings and were all crowned Eclipse champion turf male. Great-grandson Conduit (2008 and 2009) and great-great-grandson High Chaparral (2002 and 2003) each captured two consecutive editions of the race, and they were also both crowned Eclipse champion turf male. Two winners of the Filly and Mare Turf, Islington (2003) and Zagora (2012), descend maternally from Mill Reef.

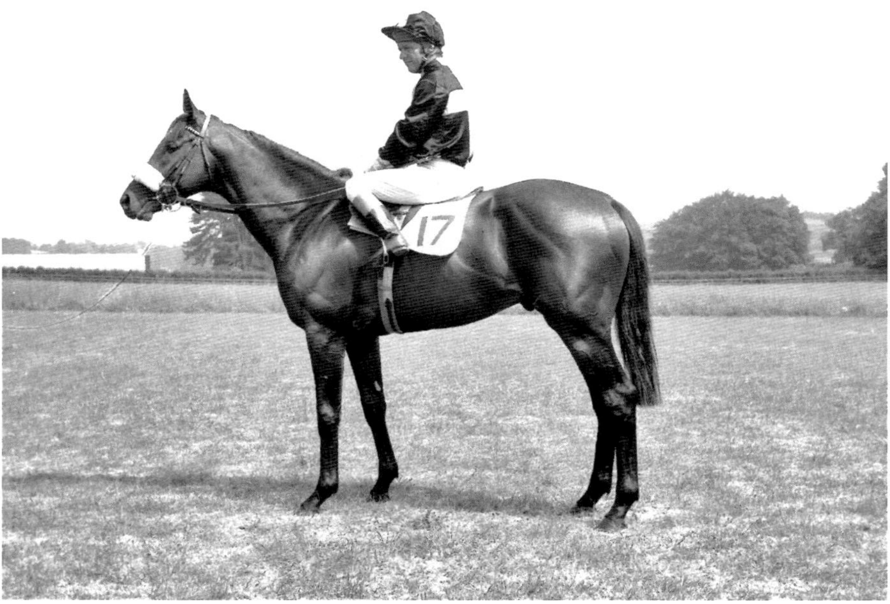

Mill Reef with jockey Geoff Lewis up, 1971. *Courtesy of Segas Picture Gallery (http://www.segaspicturegallery.co.uk/index.asp).*

Riverman was a stakes winner and leading sire in France. Sire of 128 stakes winners (13 percent), including Prix de l'Arc de Triomphe–winning fillies Detroit (1980) and Gold River (1981), he is also the damsire of champions Spinning World and Six Perfections. J.O. Tobin was champion two-year-old in England in 1976 and defeated Seattle Slew in the 1977 Swaps Stakes. He garnered co-champion sprinter honors as a four-year-old.

Other notable sons of Never Bend are American stakes-winners Proudest Roman, Never Bow and Iron Ruler. Daughters Courtly Dee and Anne Campbell were both voted Kentucky Broodmares of the Year. The latter was dam of millionaires Desert Wine and Menifee.

GREY SOVEREIGN

Grey Sovereign sired 418 foals, forty-eight of them stakes winners (10 percent). His influence as a sire was best appreciated in Europe, where he became recognized as a sire of milers and grass horses. He was champion sire of two-year-olds in England in 1958 and 1961 and in France in 1967. His son Young Emperor was crowned champion two-year-old colt in England and Ireland and became a successful broodmare sire. Stakes-winner Sovereign Path was also a leading broodmare sire and was second on the English sire list in 1970. French 2,000 Guineas–winner Zeddaan sired multiple European classics winners.

American runners Merry Ruler and Grey Monarch helped to boost Grey Sovereign's reputation as a sire in North America. But it was through his grandson Caro that Grey Sovereign established a legacy of stakes winners and leading sires in the United States. Caro was by Fortino II, a stakes winner in France, and out of the British mare Chambord. Caro sired a total of seventy-eight stakes winners (13 percent). Kentucky Derby–winning filly Winning Colors, Eclipse champion turf horse Cozzene and Canadian Triple Crown–winner, Horse of the Year and Hall of Fame inductee With Approval are among his best offspring.

Cozzene was leading sire in 1996 and sired a total of thirty-one stakes winners and fourteen millionaires, including Grey Way, champion three-year-old filly in Italy. Top California sire In Excess is by Siberian Express, another son of Caro's. In Excess's best-known son, Indian Charlie, won the Santa Anita Derby and ran third in the Kentucky Derby. He was successful at stud, siring four champions.

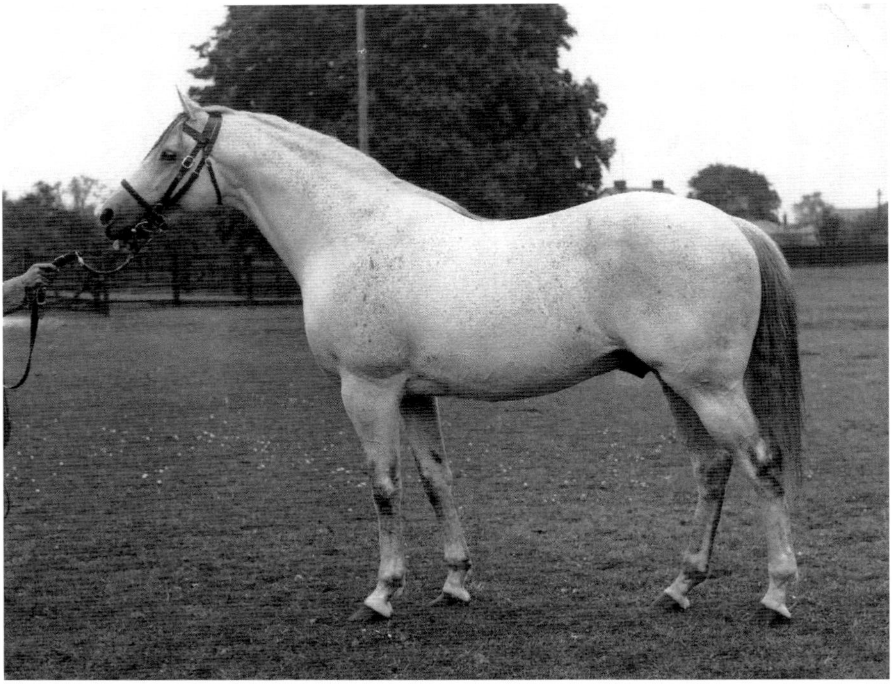

Grey Sovereign, whose descendants have had great success at the Breeder's Cup.
Courtesy of Pedlines.

The Cozzene line has had tremendous success at the Breeder's Cup. Cozzene himself captured the 1985 Breeder's Cup Mile and sired two Breeder's Cup winners in Alphabet Soup (1996 Classic) and Tikkanen (1994 Turf). Most recently, granddaughters Flotilla and Mizdirection were victorious at the 2012 Breeder's Cup at Santa Anita, capturing the Juvenile Fillies Turf and Turf Sprint, respectively. Numerous other Breeder's Cup winners trace to Caro through his other lines.

Grey Sovereign's daughters produced seventy-five stakes winners. Perhaps most notable is the mare Grey Table, dam of Quiet Little Table, which upset Horse of the Year Forego in the 1977 Suburban Handicap. Grey Sovereign also figures prominently in the bloodlines of several European Warmblood registries.

Red God

Red God stood at Loughton Stud in County Kildare, siring ten stakes winners with over £1 million in earnings, including Red Lord, Valiyar and Stanford. He is best known, however, for establishing one of Nasrullah's greatest sire lines through his son Blushing Groom. Out of the mare Runaway Bride, the chestnut horse was bred by J. McNamee Sullivan and owned by the Aga Khan. Blushing Groom was crowned champion two-year-old colt in France, ran third in the 1977 Epsom Derby and earned champion three-year-old honors in France and England. At stud, he was leading broodmare sire (1988, 1995) and leading sire (1989) in Great Britain and Ireland.

Blushing Groom led the world in stakes winners in 1985 (he sired a total of ninety-two) and is now recognized as a classic chef-de-race. He sired multiple champions, including Runaway Groom, which, unraced at two, took North America by storm as a three-year-old, placing in the Queen's Plate and winning the other two legs of the Canadian Triple Crown (the Prince of Wales Stakes and Breeder's Stakes) before going on to Saratoga to defeat all three winners of that year's American Triple Crown in the Travers Stakes. He won a Sovereign Award as Canada's champion three-year-old colt and was inducted into the Canadian Horse Racing Hall of Fame.

Blushing Groom's other champions included Nashwan (Epsom Derby and 2,000 Guineas winner and sire of Swain); Rainbow Quest (1985 English and French champion older male); Blushing John (1989 U.S. champion older male); Arazi (U.S. champion two-year-old colt and European Horse of the Year); and Sky Beauty (U.S. filly Triple Crown winner and Champion older female horse), but Blushing Groom is better known for the dynasty he established as a sire of sires and sire of broodmare sires.

Rainbow Quest led England's sire list four times and was leading broodmare sire in Great Britain and Ireland in 2003 and 2004. He sired over one hundred stakes winners and numerous classics winners. Rahy was a leading broodmare sire and also sired Hall of Fame inductee Serena's Song; two-time Horse of the Year Fantastic Light; and Mariah's Storm, the mare that was the inspiration for Soñador, the equine hero of the Dreamworks movie *Dreamer*. She is the dam of Giant's Causeway, European Horse of the Year and a leading sire in North America in 2009 and 2010.

Blushing Groom's other notable sire sons are Candy Stripes, which was sire of Horse of the World Invasor, as well as American Champion Male

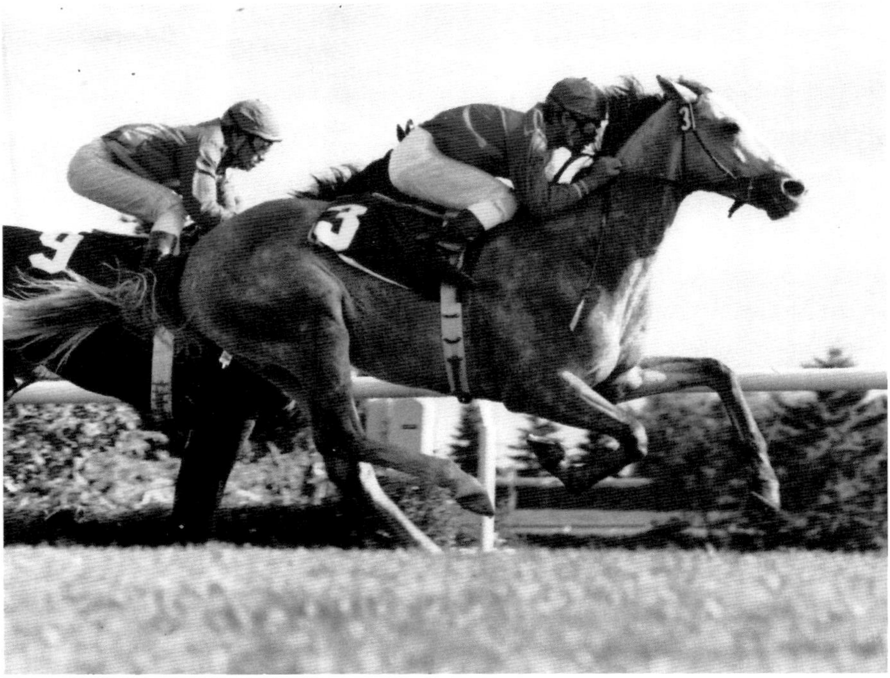

Runaway Groom, 2001 Canadian Horse Racing Hall of Fame inductee. *Courtesy of the Canadian Horse Racing Hall of Fame.*

Turf Horse Leroidesanimaux, and broodmare sire of unbeaten Candy Ride; Runaway Groom (which sired Breeder's Cup Sprint–winner Cherokee Run); and millionaires Down the Aisle, Wekiva Springs and Machikane Hokushin.

LEADING LADIES

Nasrullah established a breeding legacy not only through his sons but through his daughters as well. Dangerous Dame, which raced for Cain Hoy Stable, produced full sisters Hidden Talent and Heavenly Body (both out of Kentucky Derby–winner Dark Star), two broodmares that have "established dynasties that have lasted into the twenty-first century."[95] Hidden Talent won the Kentucky Oaks and Ashland Stakes. Dam of Kentucky Broodmare of the Year Too Bald, she is also granddam of top

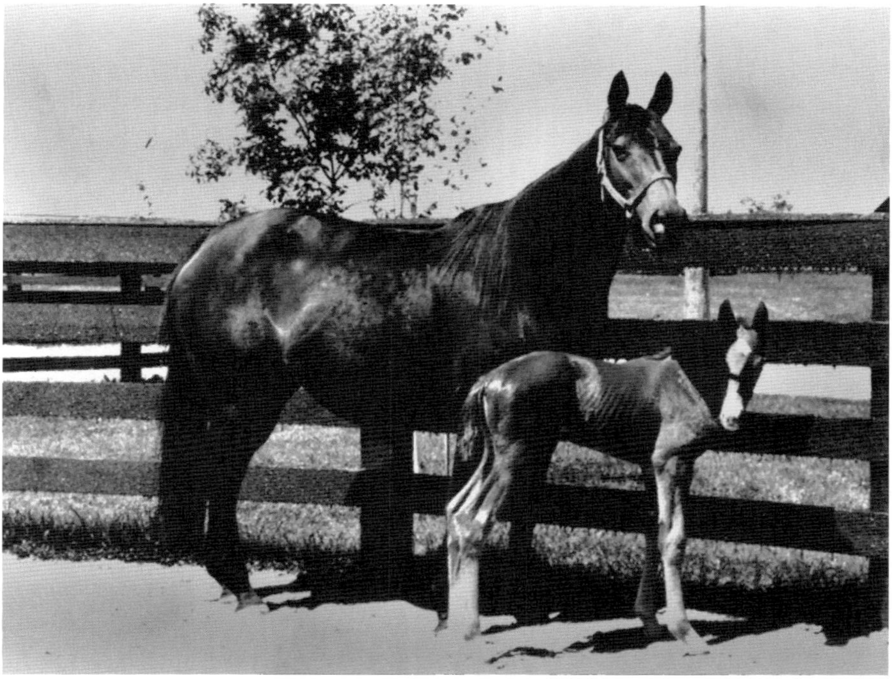

No Class, Canadian Broodmare of the Year, 1985. *Courtesy of the Canadian Horse Racing Hall of Fame.*

sires Capote, Exceller, Baldski and American Standard. Heavenly Body was a stakes winner at ages two and three. She produced stakes-winner A Thousand Stars (by Hoist the Flag) and is great-granddam of Prix de l'Arc de Triomphe–winner Sakhee.

Several of the stallion's mates, as well as many of his female descendants, have been named Kentucky Broodmare of the Year (KBOTY). Miss Disco produced six foals sired by Nasrullah, including stakes-winners Bold Ruler, Independence and Nasco, and was named KBOTY in 1958. In 1960, Siama took the award partway on account of her stakes-winning Nasrullah sons Bald Eagle and One-Eyed King. Two years later, Track Medal took the crown owing to the successes of Outing Class and Tutankhamen.

Out of the mare Bourtai, 1968 KBOTY winner Delta was from Nasrullah's first American crop of foals. A stakes winner for Claiborne Farm, she set a new track record at Churchill Downs, running six furlongs in 1:09.3 in 1956. She became known as a great producer; among her winning offspring were stakes-winners Dike (Herbager), Cabildo (Round Table), Okavango (Herbager) and Shore (Round Table).

Many second-generation descendants of Nasrullah also received KBOTY honors. The 1977 winner, Sweet Tooth, was by On-and-On, out of Plum Cake, by Ponder. Her stakes winners were Our Mims (Herbager), Alydar (Raise a Native) and Sugar and Spice (Key to the Mint). Natasha, another mare from Nasrullah's first crop, produced 1981 winner Natashka (by the Princequillo stallion and co–Horse of the Year Dedicate), a stakes winner at ages three and four. She was dam of five stakes winners: Truly Bound (In Reality); Gregorian (Graustark); Ivory Wand (Sir Ivor); Arkadina (Ribot); and Blood Royal (Ribot). Never Bend sired two winners in Courtly Dee (1983) and Anne Campbell (1999). Bald Eagle sired 1986 winner Too Bald, which also traces to Nasrullah through her dam, Hidden Talent. Weekend Surprise, 1992 winner, was by Secretariat. Between 1997 and 2011, all but two winners of the award descend from Nasrullah.[96]

Oddly enough, female descendants of Nasrullah did not really make a mark as top producers in Canada until fairly recently. For the last five years (2007–11), the winners of the Sovereign Award for Outstanding Broodmare have all descended from Nasrullah. Interestingly, three generations of Nasrullah descendants have received this award: No Class, Classy 'n Smart and Dance Smartly.

Chapter 8

THE LEGACY CONTINUES

Nasrullah's importation to the United States is considered to be one of the most monumental events in the history of the American Thoroughbred; it was ranked number twenty-three in *Horse Racing's Top 100 Moments*, ahead of Sir Gallahad III (number twenty-four) and all other stallion imports.[97] With the arrival of his offspring, he put an end to a long era of leading sire domination by imported stallions.

Beginning with Bold Ruler in 1963, the U.S. general sire list has since been led exclusively by native-bred horses; furthermore, the majority of these descend either paternally or maternally from Nasrullah. In the second half of the twentieth century, only Northern Dancer has equaled or surpassed his influence on the modern Thoroughbred.[98] That influence extends beyond the American border. In 2013, all of the seventy-four registered Thoroughbred stallions listed as standing for stud duties in Canada trace to Nasrullah.[99]

Most telling of Nasrullah's lasting influence on the modern American Thoroughbred racehorse is the fact that all twelve winners of the 2012 Eclipse Awards[100] and all twenty-seven finalists for the 2012 Sovereign Awards, including those up for Canadian Broodmare of the Year, descend from him—many of them inbred to him, with Mr. Prospector being the most dominant line through which they trace.

Abram S. Hewitt, writing in *Sire Lines*, notes that the main question about Nasrullah as a sire concerned his temperament.[101] The truth is that many high-strung horses have proven themselves as successful sires, such as St.

Simon (to which Nasrullah traces), Hastings and Fair Play, sire of the famed Man O' War. If crossed with mares with quieter temperaments, such stallions tend to produce quality offspring. In Hewitt's words, "Nasrullah needed strains with good nervous systems and some additional stamina."[102] This is evidenced by the products of Nasrullah-Princequillo crosses. Bull Hancock himself said, "With Nasrullah, who was so temperamental, we tried to find him mares with the best temperaments."[103]

The logic behind breeding to such a temperamental stallion, while hard to comprehend by some breeders, is evidenced by the legacy of winning racehorses and successful producers Nasrullah has left behind. His Highness the Aga Khan IV provides some insight into this, recalling the following:

> *All I ever heard about Nasrullah was that he was a quasi-mad horse. He was extremely difficult to handle. Later on I learnt that sometimes you have families that lose their temperament, their will to race, or just the vivacity that you need in a great athlete. One of the ways that we try to correct that is to breed to a temperamental stallion. There is no doubt that Nasrullah brought that spark. That was madness that had to be used intelligently.*[104]

And as British turf historian John Hislop put it:

> *When a high-powered pedigree gets away with it...produces a horse in which brilliance is achieved with temperamental imbalance...the product often excels both on the race course and at stud. At the same time, on the principle that you can seldom have the jam without the powder, such a pedigree often provides a two-edged sword with regards to temperament. Nasrullah is a dynamic factor in Thoroughbred breeding, but like dynamite, must be handled with care.*[105]

E.B. Johnston, owner of Fleet Nasrullah, which established his sire's legacy in California as a leading stallion, discussed at length the task of achieving that fine but delicate balance between crazy and brilliance when breeding to the Nasrullah line:

> *There's no question about it that there were a lot of sons of Nasrullah that were very difficult to control. Some of them passed too much of that onto their get. When a horse like Fleet Nasrullah, who was difficult to control, was outcrossed to the Hyperion line or the Princequillo line or some of these other lines, you seemed to retain this brilliance and be able to control it. By*

> *control it, I mean you're able to train their offspring the way you would like to.*[106]

Sport-horse breeders have described Nasrullah as a "genetic gift that requires a skillful breeder to unwrap his full potential."[107] Inbreeding to Nasrullah, while seemingly illogical, has met with great success, exemplified in such winners as Seattle Slew and Conquistador Cielo, both of which are inbred 4x4 to him. Leon Rasmussen wrote in an article for the *Daily Racing Form*:

> *As a stallion whose temperament was suspect and quite visible in a number of his get, one would expect that inbreeding to Nasrullah would be pure folly. Yet, there are so many examples to suggest that such inbreeding is more a plus than a minus.*[108]

A quirky rhyme emerged and, while nasty, stands as a testament to the wisdom of inbreeding to Nasrullah:

> *Match Nasrullah through a son,*
> *And Nasrullah through a daughter,*
> *To make it really blister,*
> *Match Nasrullah with his sister.*

Today, Nasrullah's name appears frequently in inbred pedigrees.

Nasrullah has been designated a "Brilliant" chef-de-race. The Dosage Index is a mathematical figure used by breeders of Thoroughbred racehorses and sometimes by bettors handicapping horse races to quantify a horse's ability, or inability, to negotiate the various distances at which horse races are run. It is calculated based on an analysis of the horse's pedigree. The index is compiled by noting the presence of certain influential sires, known as *chefs-de-race* (French for "chiefs of racing" or "masters of the breed") in the first four generations of a horse's pedigree. Based on what distances the progeny of the sires excelled in during their racing careers (the distance preferences displayed by the sires themselves while racing being irrelevant), each chef-de-race is placed in one or two of the following categories, or "aptitudinal groups": Brilliant, Intermediate, Classic, Solid or Professional, with Brilliant indicating that the sire's progeny fared best at very short distances and Professional indicating an ability for very long races, with the other three categories ranking along the same continuum in the aforementioned order.

According to the Aga Khan IV, it takes a number of years for a chef-de-race to emerge. "Such a horse must not only be a good stallion, but a good producer of stallions and a good sire of mares."[109] Sons of Nasrullah that also achieved this distinction include Brilliant-Intermediate Bold Ruler; Brilliant Grey Sovereign; Intermediate-Classic Nashua; Brilliant-Intermediate Never Bend; and Classic Never Say Die.

What is perhaps most notable about the descendants of Nasrullah is the diversity and versatility exhibited in their athletic ability and racing aptitude. Nasrullah's descendants have proven themselves as sprinters, stayers, middle- and classic-distance runners, handicappers, distaffers and even jumpers. In recent years, descendants of the Bold Ruler line have been shown to exhibit stellar capabilities on synthetic racing surfaces. Bold Ruler also appears in American Quarter Horse pedigrees, through 222 of his offspring, including Secretariat. Nasrullah continues to have a dominant influence in the bloodlines of international-level jumping horses. He has led the rankings in all four categories of Olympic-style sport by the Performance Horse Registry: hunter, jumper, eventing and dressage.

Descendants of Nasrullah have also been used to improve many sport-horse lines, such as registered Paints and Warmblood breeds. In the most easterly Canadian province of Newfoundland and Labrador, where I reside and home to only several hundred horses (and very few full Thoroughbreds), it was a delightful surprise to discover that horses of various bloodlines were found to trace back to Nasrullah.

The aptly named Nazrullah, whose registered name with the Jockey Club is Forty Nine Caliber, is a Canadian-bred Thoroughbred gelding that was unraced but has made a name for himself as a three-day eventer. To date, he has competed at the Intermediate and Two-Star level and was Canada Talent Squad listed in 2012. The son of Orchid's Devil and Angel's Share traces to Nasrullah both paternally and maternally through the Mr. Prospector and Boldnesian lines and shows an uncanny resemblance to the stallion, right down to his dark bay coat and the star on his forehead.

It is also worth noting the variety of Thoroughbred bloodlines with which Nasrullah mixed well, although the Nasrullah-Princequillo nick has become renowned in breeding circles. A nick is a cross of two bloodlines that produces superior results. The mating of daughters or granddaughters of Princequillo with sons of Nasrullah has been met with great success, as evidenced by Secretariat, Mill Reef, Bold Bidder and countless others. No Class, one of the greatest blue hen mares of the twentieth century, traces maternally to both stallions; her dam, Classy Quillo, was the product of

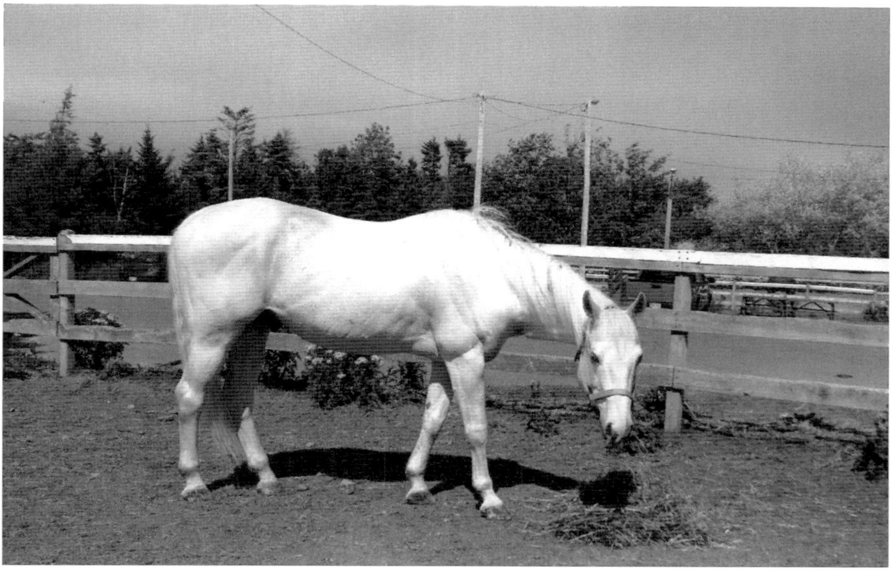

The Key Purpose, "Bob," an American Paint Horse that traces to Nasrullah through the Thoroughbred Sailor's Purpose. *Courtesy of Melanie Stone.*

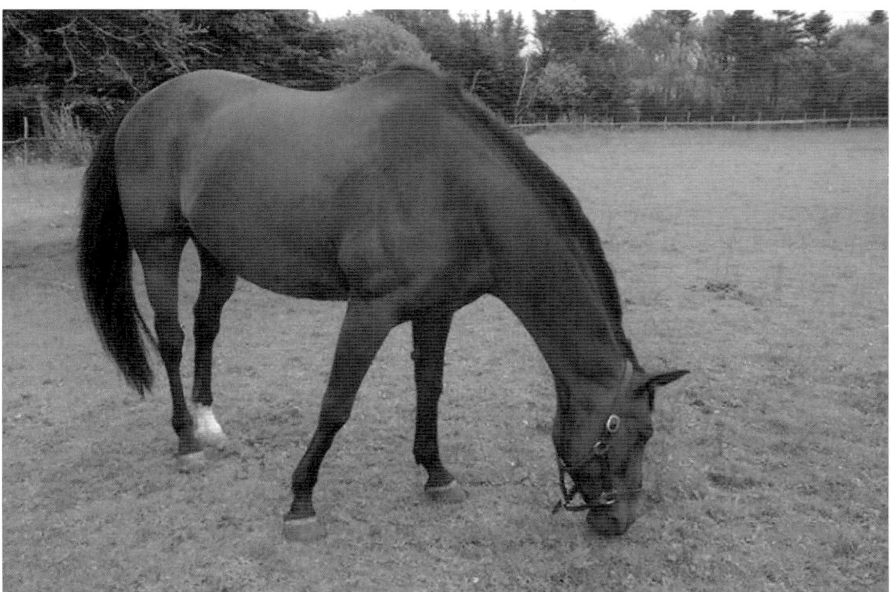

Cativa, Cativa, Cativa, a registered Canadian Sport Horse that traces to Nasrullah through the Thoroughbred Baricor. *Courtesy of Gillian Sheppard.*

mating a daughter of Princequillo (Quillopoly) with a son of Nasrullah (Outing Class). Daughters and granddaughters of Nasrullah and Bold Ruler crossed with Princequillo's son Round Table have also produced successful offspring, such as Apalachee, champion two-year-old colt in England and Ireland, and Poker, damsire of Hall of Famers Seattle Slew and Silver Charm. Despite the popularity of the Nasrullah-Princequillo nick, research has shown that Nasrullah is an even better cross with Bull Dog. Nasrullah was bred to several daughters and granddaughters of the leading broodmare sire with stellar results, including Calumet Farm's On-and-On and stakes-winning sisters Leallah and Lea Lane.

Sons and daughters of Nasrullah have also established themselves as great producers, as we have seen, extending his influence far into future generations of the breed. As we move forward through the early decades of the twenty-first century, however, that influence has become underappreciated, unrecognized and, in many cases, forgotten. It is important to remember Nasrullah's role in the shaping of the American Thoroughbred racehorse and sport-horse lines. We may never again see another like him, but his impact continues to be felt as his spirit and genes live on through horses around the world.

Chapter 9

THE HEARTS OF CHAMPIONS

While it is not uncommon for performance horses, and especially Thoroughbred breeding stallions, to succumb to ailments of the heart, particularly aortic rupture, it is worth noting that a surprising number of Nasrullah's male descendants suffered the same fate as the great stallion, dying in their late teens and early twenties of heart problems.

Impossible to ignore with regards to this is the Fleet Nasrullah branch. Beginning with Nasrullah, five generations of stallions from this line to date have died of heart attacks: Fleet Nasrullah, Gummo, Flying Paster and Flying Continental. Heart problems have also been the cause of death in stallions from the Grey Sovereign, Bold Ruler and Never Bend lines. Champion Mill Reef was humanely destroyed at the height of his stud career at age eighteen on account of what has been described as a deteriorating heart condition. His great-grandson Best Mate, a champion steeplechaser, collapsed during a race and died of a heart attack that was brought on by internal bleeding.

Other descendants of Nasrullah, including grandson Sovereign Path, great-grandsons Caro and Spectacular Bid and great-great-grandson Skywalker, all died in their twenties of heart failure. Champion Swale, a son of Seattle Slew, died of a heart attack at age three, just eight days after winning the Belmont Stakes. His death is still shrouded in mystery. Nearula died early in his stud career of a ruptured blood vessel, and interestingly, a number of horses that trace to him maternally died of aortic rupture at a young age.

Aortic ruptures, from which Nasrullah died, have been found to occur more frequently among breeding stallions of fifteen years of age or older; due to a lack of research on the causes of equine heart conditions, the possibility of a genetic predisposition to a weakened cardiovascular system cannot be confirmed.

According to equine cardiology specialist Dr. Peter W. Physick-Sheard, an associate professor in the Department of Population Medicine at the Ontario Veterinary College, University of Guelph, aortic ruptures are not an uncommon problem in older male horses, particularly racehorses. There is no evidence at this time to suggest a genetic link. The vessel ruptures at a place subject to tremendous stress. All older horses appear to be predisposed to some arterial degeneration. It appears to be a mechanical factor. It is a point of weakness and is not unique to any particular bloodline.[110]

There is reason to suspect that there may be a bigger cardiovascular contribution to sudden death in racehorses than was suspected before. Cardiac disease or cardiac malfunction might be quite important, but there is not enough data yet to confirm the extent of that importance.

Nasrullah descends from the mighty Eclipse, undefeated in eighteen races, which had a large heart. At the time of the champion racehorse's death in 1789, veterinarian Charles Vial de Sainbel performed a postmortem. He weighed the horse's heart and found that it tipped the scales at fourteen pounds, about five pounds heavier than an average horse's heart.[111]

Since that time, autopsies have revealed that many champion racehorses that descend from Eclipse had enlarged hearts, leading to the belief that these horses were "freaks," anatomically designed to perform above and beyond the capabilities of a normal horse. Australian hero Phar Lap was found to have a heart that weighed fourteen pounds, the same as Eclipse.

Based on autopsy reports, it is believed that Nasrullah's grandson Secretariat had an unusually large heart; although it was never weighed, it is speculated that it was as much as two times heavier than an average horse's heart.

According to Dr. Physick-Sheard, there is an ongoing debate over whether the human athletic heart can be considered normal or abnormal. There is certainly a lot of accumulating evidence, especially in human marathon athletes, that the more frequent and intense training, the more likely one is to acquire heart muscle problems. It could be argued that any individual who undergoes prolonged intensive training might be subject to an acquired tendency to heart problems, and of course, with underlining acquired heart disease, there is an increased chance of mortality. In the horse world, there is a similar train of thought that is currently being considered, but research is still ongoing.[112]

It is understood by equine geneticists that the large heart characteristic is inherited; it is a sex-linked gene, carried on the X chromosome, meaning that a stallion can pass the trait on only to his daughters. It is also speculated that the large heart gene is a mutation and that it should be traceable to a single ancestor. Marianna Haun is responsible for the early work on the large heart gene. Her book *The X Factor* produced the findings of a collaborative team comprising veterinarians, genetic researchers and breeders.

Recent genetic research has found that Eclipse inherited his large heart from his dam Spilletta, which carried it on both her X chromosomes, meaning she passed the trait on to all of her offspring. One of her large heart genes came from the dam line of her sire, Regulus, which traces back to Hautboy, one of the earliest Thoroughbreds and whose large heart came from his dam, a "Royal Mare." Unfortunately, as a result of poor record keeping, it has proven difficult to identify who these British- and Irish-native "Royal mares" were, as no individual names were ascribed to them in the record books. The mare Pocahontas, a descendant of Eclipse (she is line bred to him 7x7), has been identified as the principal source of the large heart gene.

It has been determined that Mahmoud, a close relative of Nasrullah, is one of the four major modern carriers of the large heart gene. Princequillo, the Irish-bred stallion and twice-leading U.S. sire whose blood nicked so well with that of Nasrullah, also carried the X factor heart and, coincidentally, died of a heart attack at twenty-four years of age. There is no established correlation between heart size and heart problems; in Dr. Physick-Sheard's opinion, a physiologically large heart does not constitute a risk factor in and of itself.

Despite its profound importance, Dr. Physick-Sheard warns against focusing so much on the heart as the sole determinant of athleticism in horses or as the source of a horse's health problems. To do so ignores the many other systems involved in the physiological response to exercise. While having a large heart is certainly an advantage, there are so many other factors to consider, and the heart itself is dependent on other systems in order to function correctly.[113]

It is clear, nonetheless, that having a physiologically large heart is a definite plus when it comes to determining the horse's maximum rate of aerobic work, but there are so many other variables at play. If a heart is large because it has pathology, however, that can cause death; cardiovascular disease can involve an enlargement of the heart. Dr. Physick-Sheard says he would like to see more research being done that focuses on the

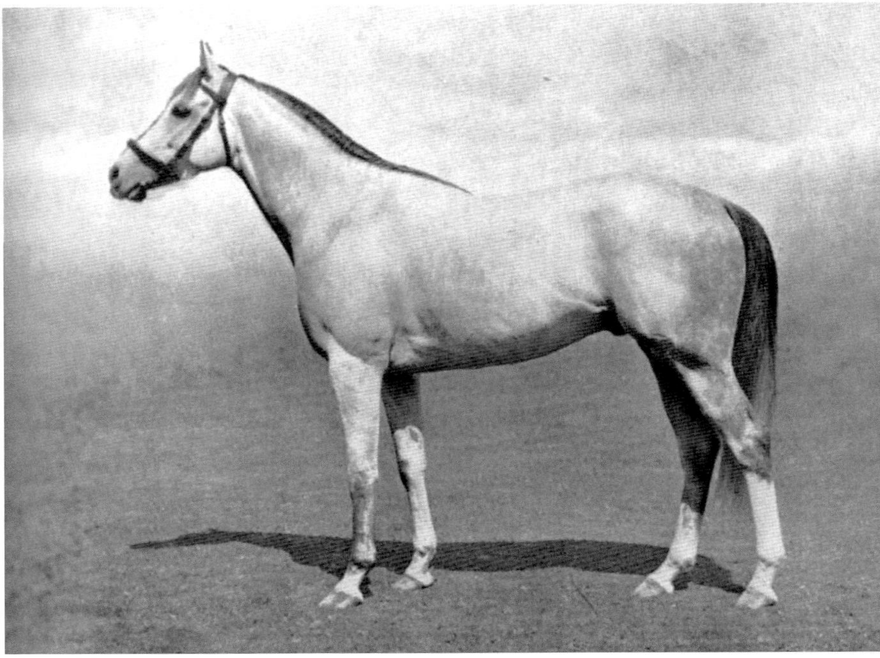

Mahmoud, one of the four major modern carriers of the large heart gene. *Courtesy of Wikipedia Commons.*

genetic contribution of the mare with regards to athletic performance, as this contribution is known to be biologically more significant than that contributed by the genes of the stallion.[114]

The equine genome has only just recently begun to be probed; future research may reveal a deeper understanding of the role of genetics in heart health and athletic performance. At the University of Guelph in Ontario, Canada, equine research on health-related topics is ongoing, including studies on arrhythmia and sudden death in racehorses. Much of this work is led by Dr. Physick-Sheard. In an attempt to better understand the causes of sudden cardiac death (SCD) and serious disturbances in heart rhythm in racehorses, heart rate and rhythm are being monitored during normal competition, from saddling to unsaddling and early recovery after the race.

Findings from a study with Standardbred racers found disturbances in rhythm after racing that would be capable of causing a fatality, but no deaths occurred in the course of the study. SCD occurs more frequently in Thoroughbreds, and atrial fibrillation, a form of arrhythmia, is the most common cardiovascular cause of poor performance in Thoroughbreds. So this breed is now the focus of further research.

It is hoped that continued research efforts will provide additional insight into the equine cardiovascular system and the causes of heart abnormalities and problematic conditions that may result in limited physical capabilities and mortality. A portion of the proceeds from the sales of this book will support graduate students conducting research on equine cardiology at the University of Guelph, Ontario Veterinary College.

Chapter 10

ECLIPSE WINNERS

Nasrullah sired nine American champions, winners of thirteen Eclipse Awards and three Hall of Fame inductees. Each horse is profiled below.

Noor

1950 Champion handicap male horse
2002 Hall of Fame inductee

Noor, whose Arabic name means "Light," was named for the famous, and supposedly cursed, Kohinoor Diamond.[115] He came from Nasrullah's first crop of foals, which arrived in 1945. Out of the mare Queen of Baghdad, daughter of British Triple Crown–winner Bahram and bred by the Aga Khan at his stud in Ireland, Noor began his racing career in England, where he spent two seasons under trainer Frank Butters (finishing third in the Epsom Derby).

In 1948, Noor was purchased in a $175,000 package with another son of Nasrullah's, Nathoo (winner of the Irish Derby), by Charles S. Howard, owner of the famed Seabiscuit, and imported to the United States. After a dismal season as a four-year-old, Noor eventually blossomed under trainer Burley Parke, becoming a successful handicapper and the archrival of American Triple Crown–winner Citation, which he defeated four times.

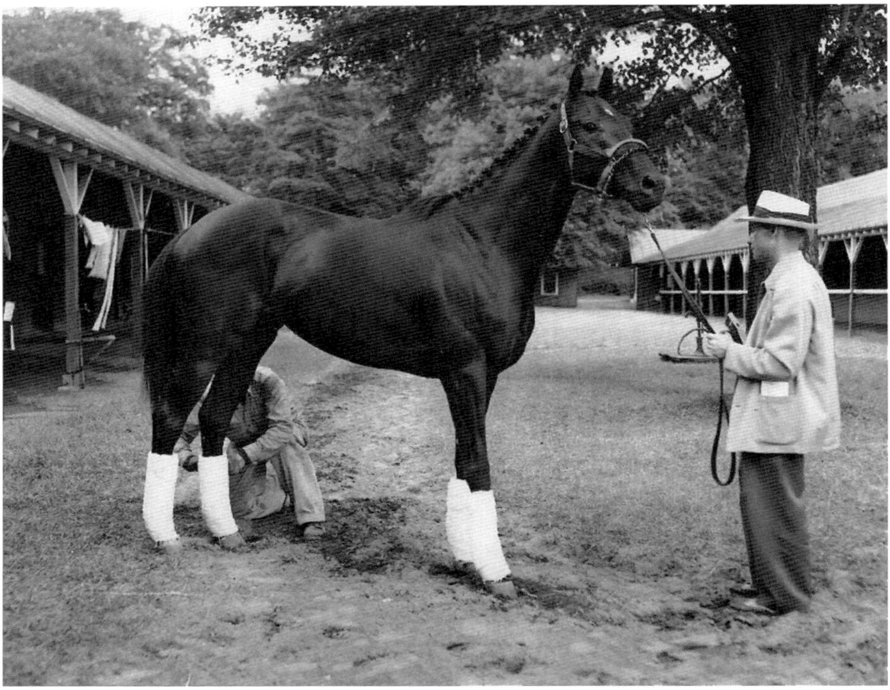

Noor, with trainer Burley Parke. *Courtesy of Gary Parke.*

He also twice defeated Assault, becoming the first horse to beat two Triple Crown winners. Noor was also the first horse to win the Big Cap (Santa Anita Handicap) and the Long Cap (Hollywood Gold Cup) in the same year (1950). In the latter, he bested Horse of the Year Hill Prince, as well as 1949 Kentucky Derby–winner Ponder and secured his place as champion handicap male horse.

Noor was retired to stud at Howard's Ridgewood Ranch at the end of his five-year-old racing season. After Charles's death, the ranch was sold, and Noor was moved first to the family's San Ysidro stables and later to Charles's son Lindsay Howard's Binglin Stable. Noor sired a total of thirteen stakes winners, including Nooran, out of the Seabiscuit mare Sea Gold; In Reserve; Prince Noor; Noor's Image; Noble Noor; Nooreddin; Flutterby; Joe Price; and Bounding Main. He was also leading broodmare sire in 1967 and 1968 and is considered to have been the most successful sire of the Howard stallions.

Noor died at the age of twenty-nine at the Loma Rica Ranch. In 2002, he was inducted posthumously into the National Museum of Racing and Hall of Fame. Recently, Charlotte Farmer led the effort to protect Noor's remains when plans were made for a business park and residential development at the site of Loma Rica Ranch. In August 2011, Noor's remains were exhumed for reburial; they were reinterred Wednesday, August 31, 2011, at Old Friends Equine in Georgetown, Kentucky. The book *Noor: A Champion Thoroughbred's Unlikely Journey from California to Kentucky* was recently published about Noor's racing career and reestablished recognition since his reburial.

Nashua

1954 Champion two-year-old colt
1955 Champion three-year-old colt and Horse of the Year
1965 Hall of Fame inductee

Out of the mare Segula, Nashua was from Nasrullah's first American crop of foals, born in 1952. Racing for Belair Stud, the bay colt was trained by Sunny Jim Fitzsimmons. Impressed by the horse's performance in a four-and-a-half-furlong maiden race, jockey Eddie Arcaro approached Fitzsimmons about gaining the mount. With Arcaro aboard, Nashua won four stakes races, including the Hopeful, Futurity and Grand Union Hotel Stakes. Six victories and two seconds earned him two-year-old male champion honors at the end of the year.

As a three-year-old, Nashua won his first time out, once again with Arcaro in the irons, taking the Spanish Moss Purse. He claimed three other titles on the Triple Crown trail, racing to victory in the Flamingo, Florida Derby and Wood Memorial. Nashua ran second to Swaps in the Kentucky Derby, a loss for which Fitzsimmons blamed himself in his underestimation of Swaps, instructing Arcaro to sit off the pace and watch out for Summer Tan.[116] Nashua retaliated with easy victories in the other two legs of the Triple Crown, setting a track record in the Preakness. That summer, he won the Arlington Classic and Dwyer Stakes. In August, he bested Swaps in a match race at Washington Park. Following a disappointing third-place finish against older horses in the Sysonby, Nashua returned to form to conclude the season with a win in the then two-mile Jockey Club Gold Cup. He was crowned champion three-year-old male and Horse of the Year.

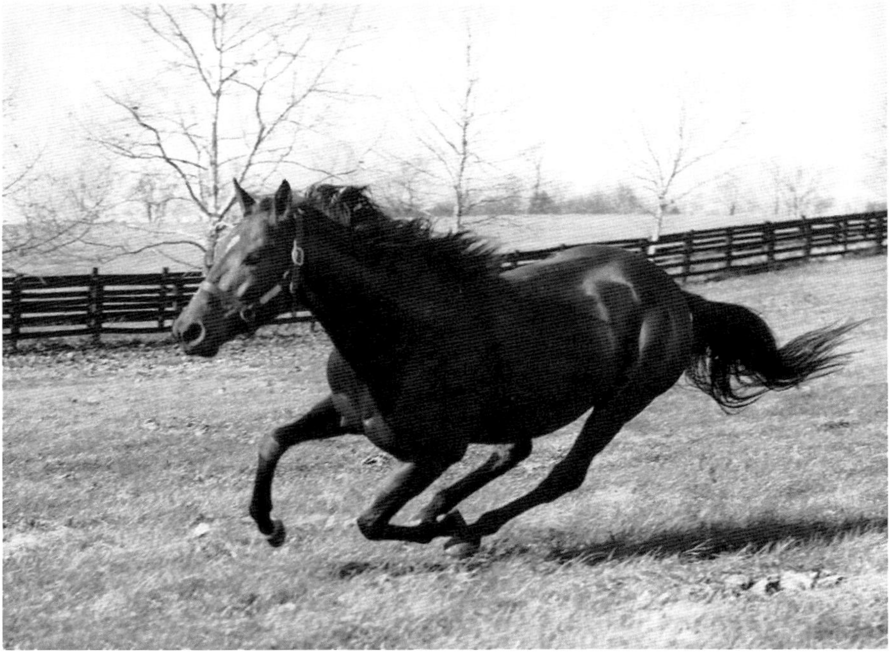

Nashua, 1956. *Courtesy of City of Bowie Museums Collection/Meadors.*

After the sudden and tragic death of owner Billy Woodward, Nashua was sold at auction at the end of 1955 for $1,252,230 to a syndicate headed by Leslie Coombs II. At four, racing for Spendthrift Farm, Nashua won six stakes races: the Widener, Grey Lag, Camden, Suburban, Monmouth Handicap and the Jockey Club Gold Cup for a second consecutive year. He retired having won twenty-two races in thirty starts, having finished out of the money only twice and with total career earnings of $1,288,565, surpassing Citation as the all-time leading money earner to date. He was inducted into the Hall of Fame in 1965.

Standing for stud duties at Spendthrift, Nashua sired 636 foals, 77 of which were stakes winners (12 percent), and he became the sixth all-time leading sire of stakes winners. He was second on the general sire list in 1964. Nashua also established himself as a top broodmare sire; his daughters produced 123 stakes winners. The Nashua Stakes, a mile-long, grade III race for two-year-olds, is run in late October each year at Aqueduct racetrack in Queens, New York.

Nasrina

1955 Co-champion two-year-old filly

Nasrina was born in 1953 and was out of the mare Red Shoes. She was bred by Bull Run Stud, owned by Howell E. Jackson and trained by Edward A. Christmas. She shared champion two-year-old filly honors with Claiborne's Doubledogdare in 1955. The Frizette and Gardenia Stakes were among her wins, and she had a third-place finish in the Selima Stakes. At four, she was only able to manage third-place finishes in the Vagrancy Handicap and New Castle Stakes. Sent to Claiborne after her racing career, and then to France in 1963, Nasrina was a dam of winners in both countries.

Bold Ruler

1957 three-year-old colt and Horse of the Year
1958 Sprinter champion
1973 Hall of Fame Inductee

Bold Ruler is arguably the most influential of Nasrullah's offspring. Out of 1958 broodmare of the year Miss Disco, to whom Nasrullah was bred six times, the bay colt was from Nasrullah's third American crop of foals, born at Claiborne in 1954.

Racing for breeder and owner Mrs. Gladys Phipps of the Wheatley Stable, Bold Ruler was trained by "Sunny Jim" Fitzsimmons and regularly ridden by Eddie Arcaro. His first stakes win came in the Youthful; in total, he had seven wins in ten races as a two-year-old, including the Juvenile, Futurity and Anticipation Purse. As a juvenile, the colt often had trouble at the gate, which might have cost him a victory on several occasions, and he gave dismal performances in his final two starts of the season, puzzling his rider and trainer.

Bold Ruler's three-year-old campaign began with a bang on January 30 at Hialeah for the Bahamas Stakes. With Ted Atkinson aboard, he won the six-furlong race by four and a half lengths. Two weeks later, he faced a narrow loss to Gen. Duke in the mile and an eighth Everglade Stakes. Bold Ruler got his revenge in the Flamingo, run at the same distance, where he defeated

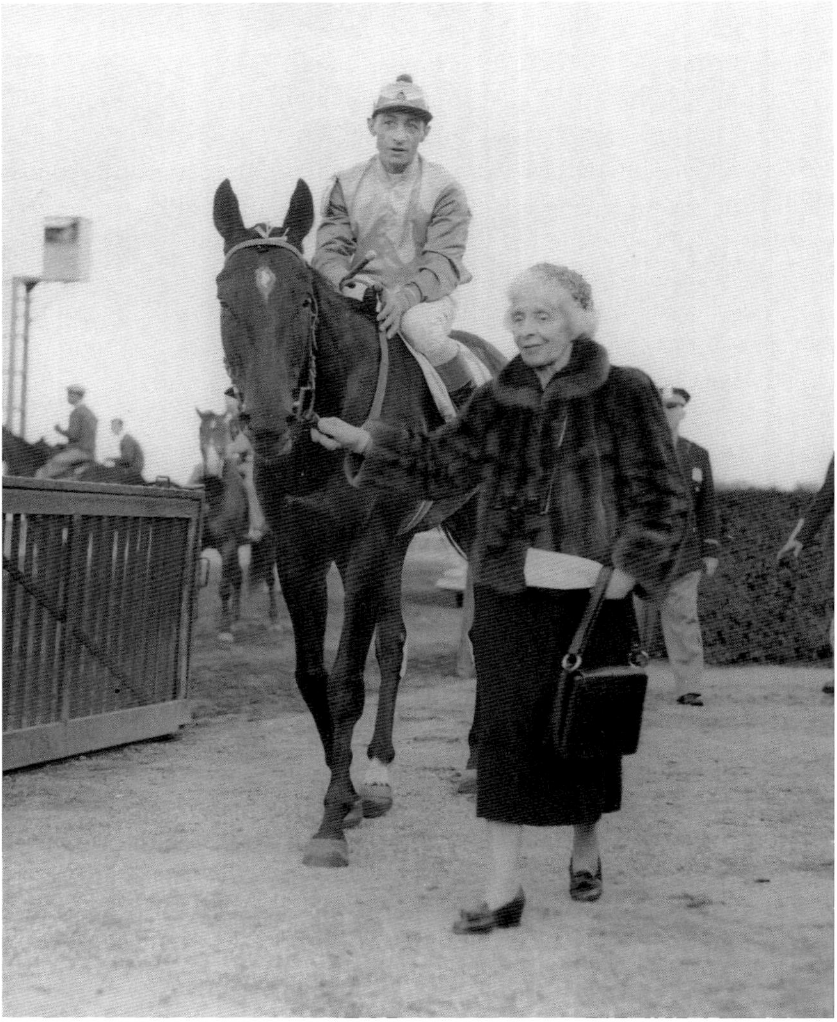

Bold Ruler with Eddie Arcaro up, being led by owner Mrs. Gladys Phipps. *Courtesy of the National Museum of Racing and Hall of Fame.*

his rival and completed the race in 1:47, the fastest time ever recorded for a three-year-old. After losing to Gen. Duke in the Florida Derby, Bold Ruler was entered as the favorite in the Kentucky Derby. Fighting Arcaro, he faded to finish fourth behind the winner, Iron Liege. That derby is remembered for jockey Willie Shoemaker's error in misjudging the finish line, costing Gallant Man the win when he stood in his stirrups and the horse slowed.

Sporting blinkers in the Preakness, Bold Ruler won by two lengths. He finished third in the Belmont, and his other wins that year included the Vosburgh, Queen's County and Trenton Handicaps, for a total of eleven races won that year. He was voted champion three-year-old colt and received Horse of the Year honors. At the end of the season, Arcaro praised the horse, saying, "For the things he can do, Bold Ruler is better than Nashua. This is a high-class horse. In fact, he's grade A."[117]

In the winter of his four-year-old season, Bold Ruler was plagued by unsoundness. He rebounded to win his first time out, which came in May for the Toboggan Handicap, and followed this up with a victory in the Carter Handicap. At year's end, he was crowned champion sprinter, having won five times at distances ranging from six furlongs to a mile and a quarter. He retired with total career earnings of $764,204.

At stud, Bold Ruler was leading sire an unprecedented eight times, surpassing Nasrullah's record of five. Eighty-two of his progeny were stakes winners (25 percent), and he produced eleven champions, including two Hall of Famers. He is best known as the sire of Triple Crown–winner Secretariat. Bold Ruler died in 1971 after undergoing treatment for cancer. He was inducted into the Hall of Fame posthumously in 1973. The Bold Ruler Handicap is a six-furlong, grade III event open to horses aged three years and up, run each October at Aqueduct Racetrack.

Leallah

1956 Co-champion two-year-old filly

Pronounced "Lee-lah," Leallah was co-champion two-year-old filly in 1956, sharing the honors with Romanita. Out of the Bull Lea mare Lea Lark, she was born in 1954 at Marchmont Farm near Paris, Kentucky. Her full sister, Lea Lane, was also a stakes winner and ran second in the 1955 Kentucky Oaks.

Bred and owned by Charleton Clay, Leallah was conditioned by Hall of Fame trainer Mackenzie Miller. She equaled the track record at Monmouth in the Colleen Stakes, completing the five and a half furlongs in 1:04.0 for the win. Other stakes wins in her two-year-old season were the Arlington Lassie, the Astoria and the Alcibiades. She also won three non-stakes races; her only loss at two came in the Princess Pat Stakes at Washington Park racetrack in Chicago, in which she finished in fourth place.

As a three-year-old, Leallah was unable to replicate her previous performances, but she did set a new track record at Keeneland, running six furlongs in 1:09, and she also won the Falls City Handicap at Churchill Downs. Out of Leallah's nine foals, three were stakes winners, the best of which was the Princequillo handicapper Go Marching, and three others were stakes placed.

NADIR

1957 Co-champion two-year-old colt

Nadir was born in 1955 and was out of the mare Gallita, a full sister to champion Gallorette. His name means "bottom," or "foundation," as it was hoped he would become a foundation stallion. Owned and bred by Claiborne Farm, the brown colt inherited his father's temperamental nature but despite this became a successful runner. He was voted co-champion two-year-old colt, sharing the honors with Jewel's Reward, after accruing five stakes wins and a second-place finish in the Cowdin Stakes. He had successful seasons at ages three and four as well.

Nadir was highly regarded by Bull Hancock, who had high hopes of winning the derby with him and rejected several offers that were made to buy the colt, including a $1 million offer made by Travis Kerr after Nadir captured the Garden State Stakes. Unfortunately, Nadir was unable to fulfill Hancock's wish and, despite being an early favorite for the derby, was withdrawn from the race, as well as the remaining Triple Crown races, after losing several prep races to eventual derby-winner Tim Tam.

Competing at shorter distances, Nadir won the six-furlong Select Handicap at Monmouth Park by a nose and the mile-and-an-eighth American Derby at Arlington Park. Running second in the Woodward and Nassau County and third in the Arlington Classic and Washington Park Handicap, he ended his three-year-old season with five wins and $204,312 in purse earnings. As a four-year-old, he won only once and finished second in the Widener Handicap. His career earnings totaled $434,316.

Retired to stand at stud at Claiborne, Nadir was only a modest success in the breeding shed. The best of his offspring to race was the top handicap horse in middle-distance races, R. Thomas. Nadir was sent to stud in Japan in 1968. In 1960, Nadir's trainer, Moody Jolley, purchased a yearling colt from Claiborne that he named Ridan (which is Nadir spelled backward), as Nasrullah's grandson reminded Jolley of the racer. Ridan would go on to be the main rival of Jaipur, another son of Nasrullah.

Bald Eagle

1960 Champion older male horse

A Claiborne-born foal of 1955, out of Acorn Stakes–winner Siama, Bald Eagle was bred by Harry F. Guggenheim. Bald Eagle was sent to England as a yearling and raced there as a two- and three-year-old. The bay horse won the Duke of Edinburgh Stakes as a two-year-old; at three, he captured the Dante and Craven Stakes races, while finishing third in the St. James's Palace Stakes. He was a beaten favorite in the 2,000 Guineas and also disappointed in the Epsom Derby.

Brought to the United States to continue his racing career for the Guggenheim's Cain Hoy Stable, Bald Eagle was placed under the command of trainer Woody Stephens. Ridden by Manuel Ycaza, who would become his regular rider, he ran third in the Massachusetts Handicap his first time out as a four-year-old and followed this up with a convincing win in the mile-and-a-quarter Suburban. Stuck in traffic in the Brooklyn Handicap, Bald Eagle finished fourth but rebounded in the Saratoga to defeat English-bred Grey Monarch (a son of Grey Sovereign) by a neck. Running second to Hillsdale in the Aqueduct Handicap and to Round Table in the Manhattan, Bald Eagle returned to form on the turf, capturing the Washington, D.C. International by over two lengths. He finished the season with a victory in the Gallant Fox handicap, setting a new track record for the mile and five-eighths, but lost out to Sword Dancer for Horse of the Year honors.

Bald Eagle started his five-year-old campaign with a second-place finish in the McLennon Handicap to half-brother On-and-On, which he went on to beat in his next two starts, the Widener and Gulfstream Park Handicaps, setting track records in both races. His other wins that year included the Metropolitan, where he set another track record, and the mile-long Aqueduct, with a second-place finish in the Man O' War and third-place finishes in the Woodward Stakes and the Jockey Club Gold Cup. Capturing the Washington, D.C. International for a second time—the first horse to do so—Bald Eagle earned the Eclipse Award for champion older male horse. In twenty-three starts, he won nine races, retiring with lifetime earnings of $676,442.

Bald Eagle was first retired to Claiborne Farm and then sent to France in 1971. While only a modest success at stud, getting just twelve stakes winners, his daughter San San won the Prix de l'Arc de Triomphe, and he is also the maternal grandsire of Breeder's Cup Juvenile winner and champion two-year-old colt Capote.

JAIPUR

1962 Champion three-year-old colt

Jaipur was a foal of 1959, out of the mare Rare Perfume, which was by the Hall of Fame stallion Eight Thirty. Jaipur was bred by Erdenheim Farms and raced for George D. Widener, winning ten of nineteen starts over a two-year race career.

Trained by Bert Mulholland and initially ridden by Eddie Arcaro, the dark bay won his first time out in a five-and-a-half-furlong sprint at Aqueduct. He also won the Flash, Hopeful and Cowdin Stakes, with second-place finishes in the Saratoga Special, Futurity and Champagne.

At three, Jaipur raced to victory in the Gotham and Withers Stakes before finishing a dismal tenth in the Preakness. He was declared the winner in the Jersey Derby after Crimson Satan was disqualified from first for interference.

Ridden by Bill Shoemaker in the Belmont, Jaipur battled with Admirals Voyage once again to get the win by a nose. He followed this up with wins in the Choice, Withers and Travers Stakes. In the latter, he ran head to head with Nasrullah's grandson Ridan in one of the most spectacular races of all time. He lost his final few races, defeated by Kelso and Ridan, but earned champion three-year-old colt honors.

Retired to stud, Jaipur sired several stakes winners, including Amber Rama, champion sprinter in England. He is also broodmare sire of 339 winners, 20 of them stakes winners.

The Jaipur Stakes, a grade III, six-furlong race for three-year-old colts run in mid-June on turf at Belmont Park, is named in his honor.

NEVER BEND

1962 Champion two-year-old colt

Never Bend was foaled at Claiborne Farm for Harry F. Guggenheim of Cain Hoy Stable in 1960 and was out of the mare Lalun, winner of the Beldame and Kentucky Oaks. A half-brother to Bold Reason, broodmare sire of Sadler's Wells and Fairy King, Never Bend was trained by Woody Stephens. At two, he triumphed in seven races, including the Futurity Stakes (besting Outing Class, another son of Nasrullah's), the Champagne Stakes and Cowdin Stakes. At year's end, he was voted U.S. champion two-year-old colt.

Never Bend's wins at three included the Flamingo Stakes and Yankee Handicap. He finished second in the Kentucky Derby behind Chateaugay and third in the Preakness. He did not run in the Belmont and placed in two other races, running second to Mongo in the United Nations Handicap and behind Kelso in the Woodward Stakes, before defeating Chateaugay by four lengths in the Yankee Handicap at Suffolk Downs. He was retired having won a total of thirteen races from twenty-three starts, earning $641,524.

Standing for stud duties at Spendthrift Farm, Never Bend was a successful sire, producing sixty-one stakes winners, and he was leading sire in Great Britain and Ireland in 1971. The most famous of his progeny was Mill Reef, winner of the Epsom Derby and Prix de l'Arc de Triomphe. Other famous progeny include Riverman, leading sire in France; world record sprinter Triple Bend; champion J.O. Tobin, which defeated Seattle Slew; and Epsom Derby–winners Shirley Heights and Reference Point.

Never Bend was also a leading broodmare sire; two of his daughters, Courtly Dee and Anne Campbell, were voted Kentucky Broodmares of the Year.

Appendix A

PEDIGREE OF NASRULLAH

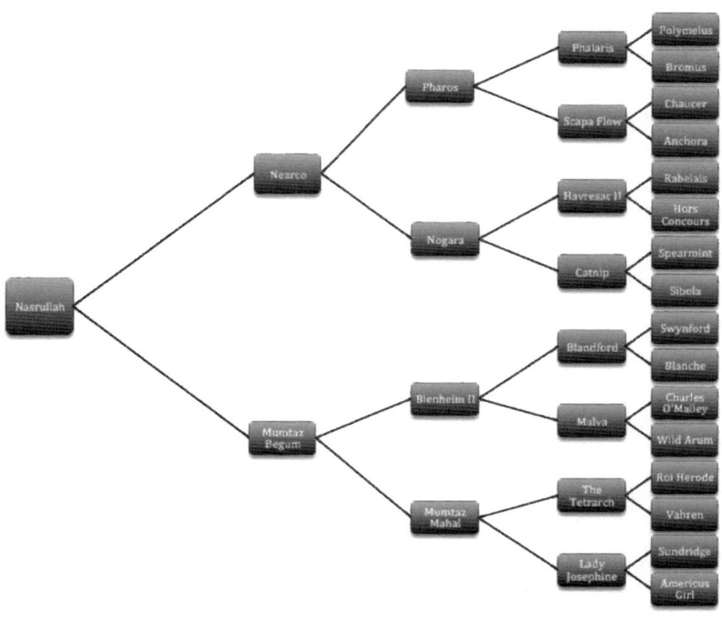

Appendix B

PROGENY OF NASRULLAH

Name	Year of Birth	Color	Dam	Dam's Sire	State Bred
Abou Ben Adhem (USA)	1960	b.	Love Game (FR)	Big Game (GB)	KY
Ace Of Trumps (USA)	1958	b.	Nance's Ace (USA)	Case Ace (USA)	KY
Adil (GB)	1950	dkb/br.	Fair Portia (GB)	Fair Trial (GB)	
Admiral Rullah (USA)	1960	b.	Lydia (USA)	Grand Slam (USA)	KY
Alarullah (USA)	1955	ch.	Marguery (USA)	Sir Gallahad (FR)	KY
Aliheim (USA)	1953	br.	Smart Niece (USA)	Alibhai (GB)	
All Ready (USA)	1957	ch.	Time To Dine (USA)	Jamestown (USA)	KY
Amanullah (USA)	1955	b.	Sun-Blest (USA)	Chance Shot (USA)	KY
Amarullah (USA)	1954	b.	Amphora (USA)	Bull Lea (USA)	KY
Ambassador (USA)*	1957	b.	Bourtai (USA)	Stimulus (USA)	KY
Anamnestes (GB)*	1948	b.	Anne of Essex (IRE)	Panorama (GB)	
Araby (GB)	1952	b.	Breide (GB)	Panorama (GB)	

Name	Year of Birth	Color	Dam	Dam's Sire	State Bred
Archery (GB)	1947	br.	Quick Arrow (FR)	Casterari (FR)	
Arctic Rullah (GB)	1954	b.	Arctic Blue (GB)	Blue Peter (GB)	
Arklow Light (GB)	1948	ch.	Novena (GB)	Tolgus (GB)	
Arullah (USA)	1960	dkb/br.	Tout a l'Heure (USA)	Eight Thirty (USA)	KY
Azarule (USA)	1957	ch.	Spring Run (USA)	Menow (USA)	KY
Baghicheh (FR)*	1951	ch.	Mehmany (GB)	Mieuxce (FR)	
Bald Eagle (USA)*	1955	b.	Siama (USA)	Tiger (USA)	KY
Ballynash (FR)*	1946	b.	Ballywellbroke (GB)	Ballyferis (GB)	
Bal Masque (USA)	1960	b.	Ballerina (USA)	Rosemont (USA)	KY
Banrullah (GB)	1946	dkb/br.	Bannlaw (GB)	Son-in-Law (GB)	
Barak (GB)	1951	b.	Flindra (GB)	Fairway (GB)	
Barnes Park (GB)*	1946	dkb/br.	Battlestown (GB)	Tolgus (GB)	
Batta (GB)	1948	br.	Roussette (GB)	Bois Roussel (FR)	
Bay Blue (USA)	1956	dkb/br.	Blue Petal (USA)	Blue Larkspur (USA)	KY
Beechpark	1949	br.	Panastrid (IRE)	Panorama (GB)	
Beguiling (USA)	1955	gr.	Dynamo (USA)	Menow (USA)	KY
Belle of All (GB)*	1948	b.	Village Beauty (GB)	Winalot (GB)	
Bellinsrullah (GB)	1950	dkb/br.	Bellinzona (GB)	Bois Roussel (FR)	
Bent Twig (USA)	1959	dkb/br.	Pines Of Rome (USA)	Roman (USA)	KY
Best Thing (USA)	1955	dkb/br.	Challadette (USA)	Challenger (GB)	KY
Billiards (USA)	1954	b.	Green Baize (USA)	Case Ace (USA)	KY
Blade of Grass (USA)	1958	b.	Blue Grass (USA)	Blue Larkspur (USA)	VA
Blair Findy (USA)	1953	ch.	Yuletide (GB)	Reveillon (GB)	KY

Name	**Year of Birth**	**Color**	**Dam**	**Dam's Sire**	**State Bred**
Blarney Castle (USA)	1954	b.	Bold Irish (USA)	Fighting Fox (USA)	KY
Blue Ruler (USA)*	1952	b.	Blue Grass (USA)	Blue Larkspur (USA)	KY
Bokara (GB)	1948	b.	Grangemore	Greek Bachelor (GB)	
Bokara (USA)	1952	ch.	Hypnotic (USA)	Hypnotist (GB)	
Bold Lark (USA)	1957	ch.	Lea Lark (USA)	Bull Lea (USA)	KY
Bold Ruler (USA)*	1954	dkb/br.	Miss Disco (USA)	Discovery (USA)	KY
Boogooloo (USA)	1954	br.	Rocking Horse (USA)	Challenger (GB)	KY
Braestar (USA)	1957	ch.	Star-Enfin (USA)	Star Pilot (USA)	KY
Brilliance (GB)	1947	dkb/br.	Lafayette	William of Valence (FR)	
Brown Sugar (USA)	1955	br.	Source Sucree (FR)	Admiral Drake (FR)	KY
Bug Brush (USA)*	1955	b.	Bonnie Beryl (USA)	Fighting Fox (USA)	KY
Bullet Hole (USA)	1957	b.	Frenabilbepa (USA)	Heliopolis (GB)	KY
Calais Dover (USA)	1956	b.	Two Cities (USA)	Johnstown (USA)	KY
Callander (GB)*	1946	br.	Calgary (GB)	Bosworth (GB)	
Cassydora (IRE)*	1948	b.	Glen Line (GB)	Blue Peter (GB)	
Castile (GB)	1947	b.	Aragon (GB)	Windsor Lad (GB)	
Catchit (IRE)	1947	b.	Cats Corrie (GB)	Bosworth (GB)	
Caval (GB)	1945	b.	Salamine (GB)	Asterus (FR)	
Ceol Srulla	1947	b.	Keyboard (GB)	Bois Roussel (FR)	
Chadwick Manor (GB)*	1948	b.	Winged Victory (GB)	Colombo (GB)	
Chesa Veglia (GB)	1959	dkb/br.	Record Serenade (GB)	Straight Deal (GB)	
Chief Scout (USA)	1954	b.	Risque Reigh (USA)	Reigh Count (USA)	KY
Chinese Map (GB)	1948	br.	Chinese Puzzle (GB)	Gainsborough (GB)	

Name	Year of Birth	Color	Dam	Dam's Sire	State Bred
Cincinnati (USA)	1954	b.	War Regalia (USA)	Man o' War (USA)	KY
Clatford (GB)	1947	ch.	Pusleigh (GB)	Bobsleigh (GB)	
Cockalorum (GB)	1945	br.	Winchmore Hill (GB)	Windsor Lad (GB)	
Cockrullah (GB)	1951	br.	Summerleaze (GB)	Fair Trial (GB)	
Cold Fury (GB)	1950	ch.	Iceland (FR)	Bobsleigh (GB)	
Colleoni (USA)	1952	b.	Winging (USA)	Sir Gallahad (FR)	
Coloullah (USA)	1953	ch.	Colosseum (USA)	Ariel (USA)	
Controlling (USA)	1959	b.	Busanda (USA)	War Admiral (USA)	KY
Coronation Boy (IRE)*	1951	b.	Persian Bloom (GB)	Furrokh Siyar (GB)	
Courtesy (USA)	1952	b.	Highway Code (GB)	Hyperion (GB)	KY
Couscous (USA)	1959	ch.	Bold Anna (USA)	Bold Venture (USA)	KY
Cumberland Gap (USA)	1956	b.	Orvieto (USA)	Challenger (GB)	KY
Curragh King (IRE)*	1950	ch.	Early Sunrise (IRE)	Sol Oriens (GB)	
Dangerous Dame (GB)	1951	ch.	Lady Kells (IRE)	His Highness (GB)	KY
Dark Caprice (GB)	1950	ch.	Madeh (GB)	Mahmoud (FR)	
Dark Days (GB)	1950	ch.	School Days (GB)	Felstead (GB)	
Dark Tiger (USA)	1959	dkb/br.	Spotted Beauty (USA)	Man o' War (USA)	VA
Daunt (USA)	1959	dkb/br.	Sunshine Nell (USA)	Sun Again (USA)	KY
Deep Blue Sea (USA)	1955	ch.	Anchors Aweigh (USA)	Devil Diver (USA)	KY
Deer Falls (USA)	1960	b.	Cascade (IRE)	Precipitation (GB)	KY
Dehav Dove (USA)	1953	b.	Black Wave (USA)	Sir Gallahad (FR)	KY
Delta (USA)*	1952	b.	Bourtai (USA)	Stimulus (USA)	KY

Name	Year of Birth	Color	Dam	Dam's Sire	State Bred
Desert King (USA)	1959	ch.	Azalea (USA)	Sun Teddy (USA)	KY
Deserted (USA)	1955	b.	Desert Sun (GB)	Hyperion (GB)	KY
Dilly King (IRE)	1947	b.	Arlington (GB)	William of Valence (FR)	
Doctor Duke (USA)	1956	b.	Effie B. (USA)	Bull Dog (FR)	KY
Dream Girl (GB)	1951	ch.	Meraline (GB)	Mieuxce (FR)	
Eastern (USA)	1952	b.	Maitresse Royale (FR)	Vatout (FR)	KY
Eastern Moonlight (GB)	1950	dkb/br.	Dangerousmoonlight (GB)	Bahram (GB)	
Eastern Princess (USA)	1958	ch.	Miss Disco (USA)	Discovery (USA)	KY
Eastern Tale (USA)	1960	b.	Belle Histoire (USA)	Blue Larkspur (USA)	KY
Eccleston Street (GB)	1949	b.	Nova Puppis (GB)	Hyperion (GB)	
Edwardsii (IRE)	1947	b.	Cattewater (GB)	Solario (GB)	
Eeyea (USA)	1951	ch.	Feale Bridge (GB)	Gold Bridge (FR)	
Eire Express (GB)	1947	b.	Das Ojos (GB)	Dastur (GB)	
El Arabi	1946	dkb/br.	Via Media (FR)	Zionist (GB)	
Ellerslie (USA)	1954	br.	Effie B. (USA)	Bull Dog (FR)	KY
Emotion	1948	b.	Boiserene (GB)	Bois Roussel (FR)	
Enagh (GB)	1948	b.	Squawking (GB)	King Salmon (GB)	
Ever More (USA)	1953	dkb/br.	Highway Code (GB)	Hyperion (GB)	KY
Explorer (USA)	1953	b.	Miss Disco (USA)	Discovery (USA)	KY
Fair Ruler (USA)	1952	b.	Fair Weather (USA)	Sickle (GB)	KY
Farullah (USA)	1954	b.	Be Faithful (USA)	Bimelech (USA)	KY
Fear Not (USA)	1960	dkb/br.	Not Afraid (USA)	Count Fleet (USA)	KY
Fiery Torch	1949	ch.	Heala Ray (GB)	Hyperion (GB)	
Final Ruling (USA)	1960	dkb/br.	Spinning Top (USA)	Bull Lea (USA)	KY
Finance (USA)	1955	br.	Busanda (USA)	War Admiral (USA)	KY

Name	Year of Birth	Color	Dam	Dam's Sire	State Bred
Firerullah (USA)*	1959	b.	Vulcania (USA)	Some Chance (USA)	KY
First Flower (USA)	1952	b.	Marguery (USA)	Sir Gallahad (FR)	KY
Fleet Marine (USA)	1960	b.	Happy Go Fleet (USA)	Count Fleet (USA)	KY
Fleet Nasrullah (USA)*	1955	dkb/br.	Happy Go Fleet (USA)	Count Fleet (USA)	KY
Flying Fury (USA)*	1952	dkb/br.	Sicily (USA)	Reaping Reward (USA)	KY
Gal I Love (USA)	1959	ch.	Gallita (USA)	Challenger (GB)	KY
Galarullah (USA)	1955	b.	Gala Belle (USA)	Sir Gallahad (FR)	KY
Galetta (GB)	1949	b.	Gally Gap (GB)	King Salmon (GB)	
Garganey (USA)	1958	b.	Lea Lark (USA)	Bull Lea (USA)	KY
Gay Nasrullah (USA)	1958	b.	Happy Go Fleet (USA)	Count Fleet (USA)	KY
Gayrullah (USA)	1959	b.	Bayadere (USA)	Chance Play (USA)	KY
Geifang Lad (GB)	1950	b.	Meraline (GB)	Mieuxce (FR)	
General Cariappa (GB)	1950	blk.	Queen of Baghdad (GB)	Bahram (GB)	
General View (GB)	1950	ch.	Distant View (IRE)	Panorama (GB)	
Gentle Ruler (USA)	1956	b.	Lithe (USA)	Pharamond (GB)	KY
Giff-Gaff (USA)	1957	br.	Source Sucree (FR)	Admiral Drake (FR)	KY
Glamour (USA)*	1953	b.	Striking (USA)	War Admiral (USA)	KY
Golestan (GB)*	1945	b.	Firanek (GB)	Cameronian (GB)	
Good Morrow (USA)	1953	b.	Sis Lea (USA)	Bull Lea (USA)	KY
Gossip Shop (GB)	1949	b.	Talk Of The Town	Baytown (IRE)	
Grand Mogul (USA)	1955	b.	Chaste Lady (USA)	Sir Gallahad (FR)	KY

NAME	YEAR OF BIRTH	COLOR	DAM	DAM'S SIRE	STATE BRED
Grey Sovereign (GB)*	1948	gr.	Kong (GB)	Baytown (IRE)	
Guide Line (USA)*	1955	ch.	Lurline B. (USA)	Alibhai (GB)	KY
Hallursan (USA)	1957	b.	Bull Poise (USA)	Bull Lea (USA)	KY
Happy Nasrullah (USA)	1959	b.	Happy Go Fleet (USA)	Count Fleet (USA)	KY
Happy Ruler (GB)	1949	br.	Happier (GB)	Flamingo (GB)	
Hard Goer (GB)	1949	ch.	Marie Galante (GB)	Brantome (FR)	
Harleigh (USA)	1956	b.	Hot Slippers (USA)	Our Boots (USA)	KY
Harvest Singing (USA)	1956	b.	Meadow (USA)	Bull Lea (USA)	KY
Hawa Bibi (GB)	1948	b.	Stella Marina (GB)	Noble Star (GB)	
Helping Hand (USA)	1953	b.	Hand Picked (USA)	Bold Venture (USA)	KY
Highland Heart (USA)	1952	b.	Misty Isle (USA)	Sickle (GB)	
Highness (USA)	1960	ch.	Miss Disco (USA)	Discovery (USA)	KY
Holiday Time (USA)	1953	b.	Picnic Lunch (USA)	Requested (USA)	KY
Homeplace (USA)	1953	b.	Blue Grass (USA)	Blue Larkspur (USA)	KY
Hot Blood (USA)	1952	ch.	Level Best (USA)	Equipoise (USA)	KY
Hurry Call (USA)	1960	ch.	Fleet Flight (USA)	Count Fleet (USA)	KY
Inaugural Day (USA)	1957	b.	Alsab's Day (USA)	Alsab (USA)	KY
Independence (USA)*	1952	b.	Miss Disco (USA)	Discovery (USA)	KY
Indian Caress (GB)	1946	b.	Pecked (GB)	Flamingo (GB)	
Indian Duel (GB)	1949	b.	Duellist	Tout Change (FR)	

Name	Year of Birth	Color	Dam	Dam's Sire	State Bred
Indian Hemp (IRE)*	1949	ch.	Sabzy (GB)	Stardust (GB)	
Indian Wave (USA)	1952	ch.	Navy Jacket (USA)	Annapolis (USA)	
Insouciant (USA)*	1952	b.	Gala Belle (USA)	Sir Gallahad (FR)	KY
Intercepted (USA)*	1959	dkb/br.	Make A Play (USA)	War Admiral (USA)	KY
Irish Abbot (GB)	1946	b.	Abbubh (GB)	Abbots Trace (IRE)	
Isaac the Angel (USA)	1957	b.	Spotted Beauty (USA)	Man o' War (USA)	KY
Island Ruler (USA)	1958	b.	Sicily (USA)	Reaping Reward (USA)	KY
Ivy (USA)	1954	b.	Hydrangea (GB)	Hyperion (GB)	KY
Jaipur (USA)*	1959	b.	Rare Perfume (USA)	Eight Thirty (USA)	KY
Jean's Joe (USA)*	1952	ch.	Pocket Edition (USA)	Roman (USA)	KY
Jedgar Ruler (USA)	1954	b.	Shy Katie (USA)	Roman (USA)	
Jeff's Uh Oh (USA)	1955	ch.	Risque Ma (USA)	Mahmoud (FR)	MD
Jerrullah (USA)^	1954	br.	Jerrybuilt (IRE)	Empire Builder (IRE)	KY
Jodhpur (USA)	1958	b.	Fleet Rose (USA)	Count Fleet (USA)	KY
Jody Belle (USA)	1953	b.	Risque Reigh (USA)	Reigh Count (USA)	KY
John Constable (GB)	1949	ch.	Painted Vale (GB)	Gainsborough (GB)	
Judy Rullah (USA)*	1953	ch.	Judy-Rae (USA)	Beau Pere (GB)	KY
Keimaneigh	1949	br.	Katushka (GB)	Colombo (GB)	
Kerne (USA)	1958	b.	Kerkeb (GB)	Straight Deal (GB)	KY
Kesrullah (GB)*	1949	ch.	Kess (FR)	Dark Legend (GB)	
Khan Of Khoristan (GB)	1949	br.	Espadrille	Mieuxce (FR)	
King's Moiety (GB)	1947	b.	Messe (GB)	Buchan (GB)	
King's Well (GB)*	1948	b.	Messe (GB)	Buchan (GB)	

Name	Year of Birth	Color	Dam	Dam's Sire	State Bred
Kingly Crown (USA)	1956	br.	Badoura (GB)	Colorado Kid (IRE)	KY
Knee Breeches (USA)	1952	b.	Fine Array (USA)	Sir Gallahad (FR)	
Lady Lune (GB)	1946	b.	La Courcelle (GB)	Knight of the Garter (GB)	
Larkswing (USA)	1959	dkb/br.	Lea Lark (USA)	Bull Lea (USA)	KY
Lashwood (GB)	1950	b.	Silver Tree (GB)	Bois Roussel (FR)	
Law And Order (USA)	1956	b.	In Bloom (USA)	Blue Larkspur (USA)	KY
Lea Lane (USA)*	1952	b.	Lea Lark (USA)	Bull Lea (USA)	KY
Lea Moon (USA)	1955	dkb/br.	Lea Lark (USA)	Bull Lea (USA)	KY
Leallah (USA)*	1954	br.	Lea Lark (USA)	Bull Lea (USA)	KY
Let's Face It (USA)	1953	b.	China Face (USA)	Display (USA)	KY
Libera Nos (GB)	1946	ch.	Malo (GB)	Mieuxce (FR)	
Liberty Ruler (USA)	1955	b.	Vagrancy (USA)	Sir Gallahad (FR)	KY
Limelight (GB)	1951	ch.	Nova Puppis (GB)	Hyperion (GB)	
Little Gidding (GB)	1949	b.	Lady Craigie (GB)	Reveillon (GB)	
Lurullah (USA)*	1957	ch.	Lurline B. (USA)	Alibhai (GB)	KY
Mafeking (GB)	1948	b.	Spion Hill (IRE)	Spion Kop (IRE)	
Magic Morn (USA)	1952	br.	Good Morning (USA)	Sir Gallahad (FR)	KY
Maid To Rule (USA)	1957	br.	Meadow Maid (USA)	Bull Lea (USA)	KY
Malka's Boy (IRE)*	1948	b.	Killiecrankie (GB)	Winalot (GB)	KY
Management (USA)	1956	b.	Busanda (USA)	War Admiral (USA)	KY
Mandil (USA)	1952	ch.	Golden Apple (USA)	Eight Thirty (USA)	KY
Manoeuvers (USA)	1952	ch.	New Weapon (USA)	Bold Venture (USA)	

Name	Year of Birth	Color	Dam	Dam's Sire	State Bred
Martin's Rullah (USA)*	1955	b.	Shy Katie (USA)	Roman (USA)	KY
Marullah (USA)*	1954	b.	Marguery (USA)	Sir Gallahad (FR)	KY
Marullo (GB)*	1946	b.	Madrigal (GB)	Prince Galahad (GB)	
Mary Be Good (GB)	1950	ch.	Marshfield (GB)	Tolgus (GB)	
Masojos (GB)	1948	b.	Das Ojos (GB)	Dastur (GB)	
Master Gunner (GB)*	1947	b.	Arsenal (GB)	Achtoi (GB)	
Me-Too (FR)	1949	ch.	Pontresina (GB)	Foxhunter (GB)	KY
Meadow Mist (USA)	1960	dkb/br.	Lea Lark (USA)	Bull Lea (USA)	KY
Megaton (USA)	1953	ch.	Nitromond (USA)	Bull Lea (USA)	KY
Metallic (GB)	1945	b.	Platinum (GB)	Sansovino (GB)	
Middletown (USA)	1955	b.	Blue Grass (USA)	Blue Larkspur (USA)	KY
Milrula (USA)	1954	dkb/br.	Miliana (FR)	Victrix (FR)	KY
Mince Pie (USA)	1954	b.	Christmastide (USA)	Pilate (USA)	KY
Miss Nasrullah (USA)	1958	ch.	Not Afraid (USA)	Count Fleet (USA)	KY
Miss Uppity (USA)*	1956	b.	Nursery School (USA)	Count Gallahad (USA)	MD
Mister Gus (USA)*	1951	b.	Fichu (GB)	Colombo (GB)	
Misty Day (USA)*	1958	gr.	Grey Flight (USA)	Mahmoud (FR)	KY
Moraine (USA)	1953	ch.	Annie Reigh (USA)	Reigh Count (USA)	KY
Musidora (IRE)*	1946	b.	Painted Vale (GB)	Gainsborough (GB)	
My Anne (GB)	1950	ch.	Court of Appeal (GB)	Apelle (ITY)	KY
Nade (USA)	1958	dkb/br.	Dentifrice (USA)	Reaping Reward (USA)	KY
Nadir (USA)*	1955	b.	Gallita (USA)	Challenger (GB)	KY
Nagea (USA)*	1957	ch.	Hydrangea (GB)	Hyperion (GB)	KY

NAME	YEAR OF BIRTH	COLOR	DAM	DAM'S SIRE	STATE BRED
Nagir (GB)	1949	ch.	Biasca (GB)	Felicitation (GB)	
Nahar (USA)	1952	b.	Chaste Lady (USA)	Sir Gallahad (FR)	
Nahodah (USA)*	1953	b.	Accra (USA)	Annapolis (USA)	
Nala (USA)*	1955	b.	Accra (USA)	Annapolis (USA)	VA
Nalanda (USA)	1959	br.	Dentifrice (USA)	Reaping Reward (USA)	KY
Nalita (USA)	1958	b.	Gallita (USA)	Challenger (GB)	KY
Nalur (USA)	1953	ch.	Lurline B. (USA)	Alibhai (GB)	KY
Namouna (USA)	1957	ch.	Virelay (USA)	Jamestown (USA)	KY
Nanka (USA)	1955	b.	Careless Nora (IRE)	Panorama (GB)	KY
Nanking (USA)	1956	b.	Challadette (USA)	Challenger (GB)	KY
Nantallah (USA)	1953	b.	Shimmer (USA)	Flares (USA)	KY
Nanticoke (USA)	1957	dkb/br.	Sea Borne (USA)	Annapolis (USA)	VA
Nantua (USA)	1954	b.	Nitromond (USA)	Bull Lea (USA)	KY
Napanee (USA)	1958	br.	Dunleith (USA)	Annapolis (USA)	VA
Narokan (USA)	1958	ch.	Mohduma (USA)	Mahmoud (FR)	KY
Nas-Mahal (USA)*	1959	br.	Love Game (FR)	Big Game (GB)	KY
Nasadear (GB)	1946	b.	Sunny Dear (GB)	Solario (GB)	
Nascania (USA)*	1957	ch.	Vulcania (USA)	Some Chance (USA)	KY
Nasco (USA)*	1955	b.	Miss Disco (USA)	Discovery (USA)	KY
Naseema (GB)	1950	br.	Evenlode (GB)	Hyperion (GB)	
Nashdara (GB)	1950	b.	Cill Dara (GB)	Bosworth (GB)	
Nashua (GB)*	1949	br.	Dasaratha (GB)	Dastur (GB)	
Nashua (USA)*	1952	b.	Segula (USA)	Johnstown (USA)	MD
Nashville (USA)*	1954	dkb/br.	Bonnie Beryl (USA)	Fighting Fox (USA)	MD
Nasomo (USA)	1956	ch.	Blue Eyed Momo (USA)	War Admiral (USA)	KY
Nasperion (USA)	1956	ch.	Hydrangea (GB)	Hyperion (GB)	KY

Name	Year of Birth	Color	Dam	Dam's Sire	State Bred
Nasr-ed-din (GB)*	1946	ch.	Chloris (GB)	Beresford (GB)	
Nasram (USA)*	1960	b.	La Mirambule (FR)	Coaraze (FR)	KY
Nasranee (USA)	1952	ch.	Bepwil (USA)	Wise Counsellor (USA)	KY
Nasreddin (USA)	1953	ch.	Appian Via (USA)	Roman (USA)	KY
Nasretta (GB)*	1950	ch.	Anne of Essex (IRE)	Panorama (GB)	
Nasrina (USA)*	1953	b.	Red Shoes (USA)	Easton (FR)	KY
Nasromond (USA)	1952	ch.	Nitromond (USA)	Bull Lea (USA)	
Nasroulette (USA)	1958	b.	Spinning Top (USA)	Bull Lea (USA)	KY
Nasrulline (USA)	1952	b.	Her Call (USA)	Case Ace (USA)	
Nassau (GB)*	1947	b.	Meraline (GB)	Mieuxce (FR)	
Nassea (GB)	1949	b.	Sea Pride (FR)	Admiral Drake (FR)	
Nassurina (USA)	1958	b.	Whirlabout (USA)	Pompey (USA)	KY
Nasturtium (GB)	1947	ch.	Chloris (GB)	Beresford (GB)	
Nasturtium (USA)	1956	b.	Blue Grass (USA)	Blue Larkspur (USA)	KY
Nasworth (GB)	1947	b.	Portworth (GB)	Portlaw (GB)	
Nasy Lady (USA)	1954	b.	Fair Weather (USA)	Sickle (GB)	KY
Natalie (USA)	1956	b.	Blue Jeans (USA)	Bull Lea (USA)	KY
Natasha (USA)	1952	b.	Vagrancy (USA)	Sir Gallahad (FR)	MD
Nathoo (GB)*	1945	gr.	Taj Shirin (GB)	Gainsborough (GB)	
Natural Bid (USA)	1957	b.	Queen of Clubs (USA)	Roman (USA)	KY
Naulahka (USA)	1952	ch.	Fairy Dream (GB)	Dastur (GB)	KY
Navan (USA)	1958	ch.	Breath Of Spring (GB)	Coup de Lyon (GB)	VA
Navarra (USA)	1953	b.	Hypnotic (USA)	Hypnotist (GB)	

Name	**Year of Birth**	**Color**	**Dam**	**Dam's Sire**	**State Bred**
Navira (USA)	1953	b.	Uvira (GB)	Umidwar (GB)	KY
Navis (USA)	1960	ch.	Sea Borne (USA)	Annapolis (USA)	VA
Nay (USA)	1959	ch.	Revoke (FR)	Big Game (GB)	VA
Nazanin (GB)	1946	b.	Carissa (GB)	Hot Night (GB)	
Nearula (IRE)*	1950	b.	Respite (IRE)	Flag of Truce (GB)	
Ned Mondrill (GB)	1949	br.	Ci Ci (GB)	Doctor Dolittle (GB)	
Nemrod	1949	b.	Theresina (GB)	Diophon (GB)	
Never Bend (USA)*	1960	dkb/br.	Lalun (USA)	Djeddah (FR)	KY
Never Say Die (USA)*	1951	ch.	Singing Grass (USA)	War Admiral (USA)	KY
New Prince (USA)	1955	b.	Princess Lea (USA)	Bull Lea (USA)	KY
New Rullah (USA)	1960	b.	War Tide (USA)	War Admiral (USA)	KY
Nirulla (USA)	1957	br.	Nitromond (USA)	Bull Lea (USA)	PA
No Angel (GB)	1949	b.	Fair Angela (GB)	Fair Trial (GB)	
Noor (GB)*	1945	br.	Queen Of Baghdad (GB)	Bahram (GB)	
Noorsong (GB)	1950	ch.	All Song (GB)	Panorama (GB)	
Nordlys (GB)	1951	gr.	Moon Drift	Mieuxce (FR)	
Not Again (USA)	1960	ch.	Tiny Request (USA)	Requested (USA)	KY
Nota Bene (GB)	1946	b.	Brasso (GB)	Beresford (GB)	
Nouronihar (GB)	1947	ch.	Sunny Maid (GB)	Lemnarchus (GB)	
Nova Herculis (GB)	1946	ch.	Smokeless (GB)	Prince Galahad (GB)	
Novarullah (IRE)*	1950	b.	Nova Puppis (GB)	Hyperion (GB)	
Novento (GB)*	1951	b.	Castagnola (GB)	Bois Roussel (FR)	
Nullabor (GB)	1950	ch.	Ortlinde (GB)	Hyperion (GB)	
Nullah (GB)	1950	ch.	Abbubh (GB)	Abbots Trace (IRE)	
Nur-ud-din (GB)	1946	b.	Valerie (GB)	Sir Cosmo (IRE)	
Nut Brown Maid (IRE)	1951	b.	Quincey (GB)	Mieuxce (FR)	

Name	Year of Birth	Color	Dam	Dam's Sire	State Bred
Nuvolari (GB)	1950	b.	Olifa (GB)	Buchan (GB)	
Nyali (USA)	1954	b.	Love Letter	Casanova (GB)	VA
Oak Cluster (USA)	1957	b.	Busanda (USA)	War Admiral (USA)	KY
On-And-On (USA)*	1956	b.	Two Lea (USA)	Bull Lea (USA)	KY
One-Eyed King (USA)*	1954	b.	Siama (USA)	Tiger (USA)	KY
Opposition (USA)	1956	b.	Antagonism (USA)	Eight Thirty (USA)	KY
Orgoglio (GB)*	1949	b.	Orienne (GB)	Sol Oriens (GB)	
Our Love (GB)	1951	br.	Glen Line (GB)	Blue Peter (GB)	
Our Nasrullah (USA)	1956	b.	Our Fleet (USA)	Count Fleet (USA)	KY
Our Rulla (USA)*	1957	dkb/br.	Shy Katie (USA)	Roman (USA)	KY
Ours (USA)	1956	b.	Evening Belle (USA)	Eight Thirty (USA)	KY
Outing Class (USA)*	1960	dkb/br.	Track Medal (USA)	Khaled (GB)	KY
Peccadillo (USA)	1952	br.	Imprudence (FR)	Canot (FR)	KY
Peep O Dawn (USA)	1957	dkb/br.	Up Early (USA)	Sun Again (USA)	KY
Penowa Rullah (USA)	1957	b.	Our Fleet (USA)	Count Fleet (USA)	KY
Perry Lad (GB)	1948	b.	Arcola (FR)	Sandwich (GB)	
Persia (USA)	1955	b.	Hydrangea (GB)	Hyperion (GB)	KY
Persian Grove (GB)	1950	ch.	Cedargrove (GB)	Tiberius (GB)	
Petticoat Rule (USA)	1957	b.	Pelisse (USA)	Pharamond (GB)	KY
Picture Gallery (GB)	1950	dkb/br.	Painted Vale (GB)	Gainsborough (GB)	
Pied D'or (USA)*	1957	b.	Two Lea (USA)	Bull Lea (USA)	KY
Pocket Ruler (USA)	1954	ch.	Pocket Edition (USA)	Roman (USA)	KY
Portrullah (GB)	1948	b.	Portworth (GB)	Portlaw (GB)	

Name	Year of Birth	Color	Dam	Dam's Sire	State Bred
Poster Girl (USA)	1960	b.	Banta (USA)	Some Chance (USA)	KY
Prince Christian (GB)	1950	b.	Church Bell (GB)	Rose Prince (FR)	
Prince Rullah (GB)	1949	b.	Novena (GB)	Tolgus (GB)	
Princely Gift (GB)*	1951	b.	Blue Gem (GB)	Blue Peter (GB)	
Pundit (USA)	1955	ch.	Time To Dine (USA)	Jamestown (USA)	KY
Punkahwallah (GB)	1949	br.	Myfan (GB)	Gainsborough (GB)	
Purdah (GB)	1946	b.	Fair City (GB)	Fairway (GB)	
Purdah (USA)	1954	b.	Uvira (GB)	Umidwar (GB)	KY
Queen Nasra (USA)	1952	b.	Bayborough (USA)	Stimulus (USA)	
Queen Of Agra (GB)	1949	b.	Queen Of Simla (FR)	Blenheim (GB)	
Racing Review (FR)	1951	br.	Merry Game (GB)	Big Game (GB)	
Rebrullah (USA)	1957	b.	Belle Rebelle (USA)	Count Fleet (USA)	KY
Red God (USA)*	1954	ch.	Spring Run (USA)	Menow (USA)	KY
Red Pants (GB)	1947	ch.	Pampenne (FR)	Pampeiro (FR)	
Regent Prince (GB)	1945	b.	Little Jess (GB)	Umidwar (GB)	
Rego (IRE)*	1949	br.	Missy Suntan (GB)	Tai Yang (GB)	
Rock Tavern (GB)	1946	ch.	Iceland (FR)	Bobsleigh (GB)	
Rocky Thumb (USA)*	1960	ch.	Skye (GB)	Blue Peter (GB)	KY
Rolling Rock (GB)	1951	br.	Salecraft (GB)	Orpen (GB)	
Rose Net (USA)*	1956	b.	Rose Fern (USA)	Fairy Manhurst (USA)	KY
Rosinante (GB)	1946	b.	Milady Rose (GB)	Knight of the Garter (GB)	
Royal Record (USA)*	1958	b.	Belle Histoire (USA)	Blue Larkspur (USA)	

Name	Year of Birth	Color	Dam	Dam's Sire	State Bred
Rue Roussel (GB)	1949	b.	Belle Roussel (GB)	Bois Roussel (FR)	
Ruffrullah (USA)	1955	ch.	Rough Shod (GB)	Gold Bridge (FR)	KY
Rule Again (USA)	1952	b.	Betagain (GB)	Gainsborough (GB)	
Ruler (USA)	1959	b.	Doubledogdare (USA)	Double Jay (USA)	KY
Rulership (USA)	1955	br.	Not Afraid (USA)	Count Fleet (USA)	KY
Rules In Air (USA)	1955	ch.	Fleet Jackie (USA)	Count Fleet (USA)	KY
Ruling Beauty (USA)	1956	b.	Spotted Beauty (USA)	Man o' War (USA)	KY
Rullah Good (USA)	1960	ch.	Case Goods (USA)	Case Ace (USA)	KY
Rullahlea (USA)	1956	ch.	Leading Home (USA)	Bull Lea (USA)	KY
Rullian (USA)	1958	dkb/br.	Brownian (USA)	Sickle (GB)	KY
Rulling (USA)	1955	ch.	Seaton Pippin (USA)	Johnstown (USA)	
Run and Rule (USA)	1956	ch.	Gallita (USA)	Challenger (GB)	KY
Russ-Marie (USA)	1956	b.	Marguery (USA)	Sir Gallahad (FR)	KY
Russian Roulette (USA)	1957	b.	Challadette (USA)	Challenger (GB)	KY
Sarah Madeline (GB)	1948	ch.	Marshfield (GB)	Tolgus (GB)	
Selenia (GB)	1949	b.	Black Selene (GB)	Hyperion (GB)	
Seminole (USA)	1959	b.	Banish (USA)	Count Gallahad (USA)	
Sepoy (GB)	1950	b.	Tropical Moon (GB)	Colombo (GB)	
She Alone (USA)	1958	dkb/br.	Lalun (USA)	Djeddah (FR)	KY
Sir Ruler (USA)*	1955	b.	My Emma (USA)	Isolater (GB)	KY
Skeeter (GB)	1951	ch.	Paravane (GB)	Fairhaven (GB)	

Name	Year of Birth	Color	Dam	Dam's Sire	State Bred
Slamruler (USA)	1957	br.	Lydia (USA)	Grand Slam (USA)	KY
Slipperoo (USA)	1954	b.	Hot Slippers (USA)	Our Boots (USA)	KY
Smarty Smarty (USA)	1952	b.	Smart Niece (USA)	Alibhai (GB)	
So Chic (USA)	1954	b.	Striking (USA)	War Admiral (USA)	KY
Somali (GB)	1945	b.	Sonibai (GB)	Solario (GB)	
Sona (USA)	1949	ch.	Golden Meadow	Gold Bridge (FR)	
South Carolina (GB)	1951	br.	Lafayette	William of Valence (FR)	
Spectrum (IRE)	1948	ch.	Rainbow Room (GB)	Precipitation (GB)	
Spinnaker (USA)	1960	b.	Admirals Belle (USA)	War Admiral (USA)	KY
St Peter's Gate (GB)	1950	ch.	Seraphic (GB)	Fair Trial (GB)	
Star of Burghley (GB)	1948	ch.	Ortlinde (GB)	Hyperion (GB)	
Starullah	1948	b.	Nova Puppis (GB)	Hyperion (GB)	
Stavroula (USA)	1956	b.	Segula (USA)	Johnstown (USA)	KY
Step On (USA)	1959	b.	Stepping Stone (USA)	Princequillo (IRE)	KY
Stev-Rullah (USA)	1958	gr.	Gay Darling (USA)	Mahmoud (FR)	KY
Stony Point (GB)	1948	b.	Pampenne (FR)	Pampeiro (FR)	
Sugar Bowl (GB)*	1947	b.	Village Beauty (GB)	Winalot (GB)	
Sun Magic (USA)	1959	b.	Ruddy (USA)	Devil Diver (USA)	KY
Sun Ruler (GB)	1951	b.	Solisequious (IRE)	Solario (GB)	
Sunny Nook (GB)	1947	b.	Sunny Dear (GB)	Solario (GB)	
Sunranee (GB)	1948	b.	Sunny Spot (GB)	Felstead (GB)	
Svelte (USA)	1957	b.	Lithe (USA)	Pharamond (GB)	KY

Name	**Year of Birth**	**Color**	**Dam**	**Dam's Sire**	**State Bred**
Sweet Pepper (FR)	1947	gr.	Saucy Bella (GB)	Bellacose (GB)	
Sweet Seven (GB)	1951	br.	Loaded Dice (GB)	Monument (IRE)	
Swiftly Morgan (USA)	1960	b.	Countess Fleet (USA)	Count Fleet (USA)	KY
Taffeta*	1949	ch.	Silver Hook (GB)	Tolgus (GB)	
Talbot's Inch (GB)	1947	ch.	Sunny Spot (GB)	Felstead (GB)	
Tambara (GB)*	1947	ch.	Theresina (GB)	Diophon (GB)	
The Irishman (USA)*	1956	br.	Bold Irish (USA)	Fighting Fox (USA)	KY
The Twister (GB)	1946	b.	Clairvoyante (FR)	Clarissimus (GB)	
Theophilus (GB)	1951	br.	Temair (GB)	Stardust (GB)	
Thicket (GB)	1947	b.	Thorn Wood (GB)	Bois Roussel (FR)	
Tico Tico	1947	br.	River Princess (GB)	Windsor Lad (GB)	
Tillman (USA)	1958	gr.	Spotted Beauty (USA)	Man o' War (USA)	KY
Tinker Toy (USA)	1953	ch.	Tocada	The Druid (GB)	KY
Tinto (GB)	1948	b.	Gold Maid (GB)	Poor Man (GB)	
Top Of The Morning (GB)	1949	br.	Lovely Day (GB)	Dastur (GB)	
Triton (GB)	1951	br.	Sea Pride (FR)	Admiral Drake (FR)	
Tuffley Lad (GB)	1949	ch.	Battle On	Coup de Lyon (GB)	
Tulsan (USA)	1954	br.	In Bloom (USA)	Blue Larkspur (USA)	KY
Turtle Dove (USA)	1958	ch.	Fond Embrace (USA)	Haltal (USA)	KY
Tutankhamen (USA)*	1958	br.	Track Medal (USA)	Khaled (GB)	KY
Tynedale (GB)	1947	b.	Gold Maid (GB)	Poor Man (GB)	
Udaipur (USA)	1957	dkb/ br.	Rare Perfume (USA)	Eight Thirty (USA)	KY
Urshalim (GB)*	1951	br.	Horama (GB)	Panorama (GB)	
Valerullah (GB)	1951	ch.	Painted Vale (GB)	Gainsborough (GB)	

NAME	YEAR OF BIRTH	COLOR	DAM	DAM'S SIRE	STATE BRED
Vicerullah (USA)	1955	b.	Strange Device (USA)	Stimulus (USA)	KY
Victory Morn (USA)*	1955	dkb/ br.	Good Morning (USA)	Sir Gallahad (FR)	KY
Victory Roll (GB)*	1950	b.	Chinese Puzzle (GB)	Gainsborough (GB)	
Vulcan Rullah (USA)	1958	b.	Vulcania (USA)	Some Chance (USA)	KY
Wagaway (USA)	1954	b.	Tocada	The Druid (GB)	KY
War Plume (USA)	1954	b.	Briarhone (USA)	Sir Gallahad (FR)	KY
Warlike (USA)	1957	b.	Antagonism (USA)	Eight Thirty (USA)	KY
Warrens Pride (GB)	1946	br.	Das Ojos (GB)	Dastur (GB)	
Waterhall (GB)*	1950	b.	Estuary (GB)	Orwell (GB)	
Way We Go (USA)	1957	b.	Ballerina (USA)	Rosemont (USA)	KY
Wayward Miss (USA)	1956	br.	Imprudence (FR)	Canot (FR)	KY
We Don't Know (IRE)	1950	b.	Marie Galante (GB)	Brantome (FR)	
Whistling Wind (USA)	1959	b.	Spring Run (USA)	Menow (USA)	KY
Windsor Ruler (USA)	1956	b.	Windsor Whisper (GB)	Windsor Slipper (GB)	KY
Winged Victory (USA)	1958	ch.	Java Belle (USA)	Djeddah (FR)	KY
Without Fail (USA)*	1960	br.	Dentifrice (USA)	Reaping Reward (USA)	KY
World Traveler (USA)	1957	dkb/ br.	Bold Anna (USA)	Bold Venture (USA)	KY
Worthy Effort (GB)	1950	b.	Carpet Slipper (IRE)	Phalaris (GB)	
Zarin Taj (GB)	1945	b.	Parmys (GB)	Fairway (GB)	
Zonah (USA)	1958	b.	Gambetta (USA)	My Babu (FR)	KY
Zucchero (GB)*	1948	br.	Castagnola (GB)	Bois Roussel (FR)	

* Stakes winner

^ One of a twin; the other foal died

NOTES

CHAPTER 1

1. Jodidio, *Racing and Breeding Tradition*, 108.
2. Aga Khan III, *Memoirs,* http://ismaili.net/heritage/files/The%20Memoirs%20of%20AGA%20KHAN-1.pdf.
3. Jodidio, *Racing and Breeding Tradition*, 43.
4. Ibid., 41.
5. Ibid., 48–49.
6. Aga Khan III, *Memoirs*.

CHAPTER 2

7. Americus was a stakes winner in the United States, where he raced as Rey del Caredes (1894) and Rey del Carreres (1895). He was exported to England, where his name was changed to Americus; he was a stakes winner there as well.
8. Edwards, *Racehorse*, 72.
9. Thoroughbred Heritage, "The Tetrarch," http://www.tbheritage.com/Portraits/TheTetrarch.html.
10. Ibid., "Blenheim," http://www.tbheritage.com/Portraits/Blenheim.html. According to Hewitt in *Sire Lines*, Blenheim was sold for $250,000.
11. Lennox, *Dark Horse*, 97–98.

12. Ibid.
13. Spearmint sired Plucky Liege, one of the best broodmares of the twentieth century, dam of leading sires Sir Gallahad III and Bull Dog.
14. Thoroughbred Heritage, "Pharos," http://www.tbheritage.com/Portraits/Pharos.html.
15. Phalaris was the leading sire of two-year-olds in 1925, 1926 and 1927, leading overall sire twice, in 1925 (the year Manna won the 2,000 Guineas and Derby Stakes) and 1928 (the year Fairway was a three-year-old). He was also second on the list in 1926, third in 1927 and fourth in 1930. He was leading broodmare sire in 1937, 1941 and 1942.
16. Thoroughbred Heritage, "Phalaris," http://www.tbheritage.com/Portraits/Phalaris.html.
17. Cyllene sired four Derby winners: Cicero (1905), Minoru (1909), Lemberg (1910) Tagalie (1912).

CHAPTER 3

18. Aga Khan III, *Memoirs*, 212.
19. Ibid.
20. Hewitt, *Sire Lines*, 143.
21. Capps, *Spectacular Bid*, 20.
22. Heather Bailey, "Nearco, Nasrullah and Royal Charger: A Triptych Of Three-Day Breeding," *Chronicle of the Horse*, http://www.chronofhorse.com/forum/archive/index.php/t-32754.html.
23. Mitchell, *Great Breeder's*, 21.
24. Richards, *My Story*, 145.
25. Whitney Tower, "Nashua's Sire and Mr. Fitz." *Sports Illustrated*, November 1, 1954, 44, http://sportsillustrated.cnn.com/vault/article/magazine/MAG1128737/index.htm
26. Richards, *My Story*, 145.
27. During both world wars, the race was transferred to Newmarket from Epsom Downs.
28. A weight-based assessment of the previous year's two-year-olds based on a theoretical race at $1^1/_{16}$ miles.
29. Bowen, *Dynasties*, 120.
30. Hewitt, *Sire Lines*, 140.
31. Phil Bull, *Best Horses of 1943*, 153.

32. Richards, *My Story*, 146.
33. Aga Khan III, *Memoirs.*
34. Between 1942 and 1944, the race was referred to as the New St. Leger Stakes and was transferred from Doncaster racecourse to Newmarket (Summer Course).
35. There is some discrepancy over Nasrullah's total career earnings. According to Bowen in *Dynasties*, it was $15,217 (p. 120); in *Nashua*, he cites $15,240 (p. 54). Whitney Tower, in "Nashua's Sire and Mr. Fitz," (p.44) reports $15,263.
36. Bull, *Best Horses of 1943*, 156.
37. Ibid.

CHAPTER 4

38. Bull, *Best Horses of 1943*, 156.
39. Toby, *Noor*, 22.
40. Tower, "Nashua's Sire and Mr. Fitz," 44.
41. Morris and Randall, *Century of Champions*.
42. Roger Mortimer, Richard Onslow and Peter Willett, *Biographical Encyclopedia*.
43. *Irish Horse*; Abelson and Tyrrel, *Breedon Book of Horse Racing Records*.
44. Morris and Randall, *Horse Racing*; Mitchell, *Great Breeders.*
45. *Calgary Herald*, "Belle of All Wins Guineas," May 4, 1951.
46. "Neasham Belle Best in Oaks," *Leader-Post*, June 2, 1951.
47. Mortimer, Onslow and Willett, *Biographical Encyclopedia.*
48. John Randall, "Supreme Court First to Secure a Verdict," *Racing Post*, July 21, 2008.
49. Mortimer, Onslow and Willett, *Biographical Encyclopedia*.
50. Bull, *Racehorses of 1953*, as cited by John Randall, "Original Zucchero as Difficult as He Was Brilliant," *Racing Post*, March 25, 2002, http://www.thefreelibrary.com/Original+Zucchero+as+difficult+as+he+was+brilliant.-a084132115
51. Randall, "Original Zucchero."
52. Bowen, *Matriarchs Volume II,* 35.
53. Richards, *My Story*, 146.

CHAPTER 5

54. John P. Sparkman, "History of Leading Sires of North America," *Thoroughbred Times*, http://www.thoroughbredtimes.com/stallion-directory/media/resources/history-of-leading-sires-in-north-america.pdf.
55. Ibid. Revenue was the first recorded leading North American sire by earnings, represented by ten starters, forty-six wins and total earnings of $49,450.
56. The syndicate that purchased Blenheim II consisted of Arthur B. Hancock Sr., Warren Wright of Calumet Farm, William DuPont, Mrs. Thomas H. Somerville, J.H. Whitney of Greentree Stud, John D. Hertz of Stoner Creek Stud and Robert A. Fairbairn of Fairholme Farm. Jodidio, *Racing and Breeding Tradition*, 5.
57. Tony Sweeney, http://www.agakhanstuds.cm/history/ch6.asp, as cited by Jodidio, *Racing and Breeding Tradition*, 60.
58. This quote, which originally appeared in a 1971 edition of the *Blood-Horse*, was cited by Bowen, in *Dynasties*, 117, as well as in *Legends of the Turf*, 253.
59. Kentucky Thoroughbred Association, Inc., and Kentucky Thoroughbred Owners and Breeders, Inc.
60. Bowen, *Legends of the Turf*, 63.
61. Whitney Tower, "Biggest Bull in the Market," *Sports Illustrated*, August 8, 1960, 30. http://sportsillustrated.cnn.com/vault/article/magazine/MAG1071643/index.htm#
62. Hancock was joined by Robert A. Fairburn, William Woodward Sr. and Marshall Field III in forming the first modern stallion syndicate. They paid $140,000 for Sir Gallahad III. Glenye Cain Oakford, "Claiborne at 100," *Daily Racing Form*, July 31, 2010, http://www.drf.com/news/claiborne-farm-100
63. In exchange for standing Princequillo at Ellerslie Stud, the Hancocks received a quarter-interest in him. He was later sent to Claiborne Farm after Ellerslie was sold.
64. The syndicate that purchased Nasrullah also included John D. Hertz, J.S. Phipps, Henry Carnegie Phipps of Wheatley Stable, Marion du Pont Scott, George D. Widener of Erdenheim Farm, Charlton Clay, Howell E. Jackson, J.T. de Blois Wack, Dr. Eslie Asbury, Thomas M. Girdler and Clifford Mooers. David Schmitz, "Fifty Years of Nasrullah," *Blood-Horse*, July 9, 2001, http://www.bloodhorse.com/horse-racing/articles/4765/fifty-years-of---nasrullah
65. Bowen, *Dynasties*, 117.

66. Neil McCarthy purchased the Australian runner Shannon in 1948, winner of the Hollywood Gold Cup and voted Eclipse champion older male. He also imported the Irish-bred Ole Fols, a stakes winner from the ages two to six in the United States.
67. There is some discrepancy over the exact figure that Nasrullah was syndicated for. Hancock reportedly quoted this figure in *International Stallions and Studs*, as cited by Bowen, in *Dynasties*, 117, and in *Horse Racing's Top 100 Moments*, 95. Articles in *Thoroughbred Record*, May 30, 1959, 22, and *Daily Racing Form*, September 29, 1986, both report it was $372,000, while the *Blood-Horse*, May 30, 1959, 1020, claims it was $370,000.
68. Mitchell, *Great Breeders*, 21.

Chapter 6

69. *Canberra Times*, July 5, 1950, http://trove.nla.gov.au/ndp/del/article/2787159/696163?zoomLevel=3.
70. The video of Nasrullah arriving at Claiborne can be viewed at the following link: http://link.brightcove.com/services/player/bcpid13320747001?bctid=89522948001.
71. Mitchell, *Great Breeders*, 30.
72. Bowen, *Dynasties*, 121.
73. Bailey, "Nearco, Nasrullah and Royal Charger."
74. Tracey Gantz, "Fire in the Pedigree," *Thoroughbred of California*, October 1983, 107.
75. Bailey, "Nearco, Nasrullah and Royal Charger."
76. Tower, "Nashua's Sire and Mr. Fitz," 66.
77. Hollingsworth, *Kentucky Thoroughbred*, 144.
78. Mitchell, *Great Breeders*, 30.
79. Stud News, *Blood-Horse*, May 30, 1959, 1020.
80. *Thoroughbred Record*, "Nasrullah: March 2, 1940–May 26, 1959," May 30, 1959, 22.
81. Whitney Tower, "Nashua: Horse to Watch," *Sports Illustrated*, October 18, 1954, 27, http://sportsillustrated.cnn.com/vault/article/magazine/MAG1128803/index.htm.
82. Mitchell, *Great Breeders*, 31.
83. *Blood-Horse*, 1965, interview with Ellwood Johnston, http://www.secondrunning.com/Fleet%20Nasrullah.htm.
84. Kent Hollingsworth, "Nasrullah Death Unexpected in Claiborne Farm Paddock," *Lexington Leader*, May 26, 1959.

85. Nack, *Secretariat*, 59.
86. Tower, "Biggest Bull in the Market," 33.
87. Stud News, *The Blood-Horse*, May 30, 1959, 1020.

CHAPTER 7

88. Bailey, "Nearco, Nasrullah and Royal Charger."
89. Hollingsworth, "Nasrullah Death Unexpected."
90. Tower, "Biggest Bull in the Market," 30.
91. Mary Simon, "A Century of Excellence," *Thoroughbred Times*, September 11, 2010, 11.
92. As of 2012, Riva Ridge, Affirmed, Sunny's Halo, Ferdinand and Sea Hero are the only Kentucky Derby winners who do not descend from Nasrullah.
93. John P. Sparkman, "Nasrullah in the 21st Century," *Thoroughbred Times*, January 20, 2007, 26.
94. The first of the two colts to be named Bold Lad, born in 1964, was Irish bred and was crowned champion two-year-old colt in England and Ireland.
95. Bowen, *Matriarchs Volume II*, Dangerous Dame photo insert.
96. The only two Kentucky Broodmare of the Year winners between 1997 and 2011 that do not descend from Nasrullah are Toussaud (2002) and Baby Zip (2005).

CHAPTER 8

97. *Horse Racing's Top 100 Moments.*
98. John P. Sparkman, "Marvelous Marguerite's Line," *Thoroughbred Times*, April 21, 2001, http://www.thoroughbredtimes.com/weekly-feature-articles/2001/april/21/marvelous-marguerites-line.aspx
99. 2013 Sire Book, *Canadian Thoroughbred.* Information compiled by the National Office of the Canadian Thoroughbred Horse Society.
100. Wise Dan took home three Eclipse Awards: Horse of the Year, older male and male turf horse.
101. Hewitt, *Sire Lines*, 34.
102. Ibid, 34.
103. Tower, "Biggest Bull in the Market," 34.

104. Jodidio, *Racing and Breeding Tradition*, 216.
105. Bailey, "Nearco, Nasrullah and Royal Charger."
106. Tracey Gantz, "Fire in the Pedigree," *Thoroughbred of California*, October 1983, 111.
107. http://www.sport-horse-breeder.com/nasrullah.html.
108. Leon Rasmussen, "Nasrullah America's Gain, England's Loss," *Daily Racing Form*, September 29, 1986.
109. Jodidio, *Racing and Breeding Tradition*, 164.

CHAPTER 9

110. Interview with Dr. Peter Physick-Sheard, November 28, 2012.
111. Clee, *Eclipse*, 187.
112. Interview with Dr. Peter Physick-Sheard.
113. Ibid.
114. Ibid.

CHAPTER 10

115. Toby, *Noor*, 24.
116. *Sports Illustrated*, May 2, 1955, as cited in Gatto, *Belair Stud*, 92.
117. Bowen, *Bold Ruler*, 129.

SOURCES

Abelson, Edward, and John Tyrrel. *The Breedon Book of Horse Racing Records*. Breedon Books Publishing, 1993.

Aga Khan III. *The Memoirs of the Aga Khan, World Enough and Time*. New York: Simon and Schuster, 1954.

Blood-Horse

Bowen, Edward L. *Dynasties: Great Thoroughbred Stallions*. Lexington, KY: Eclipse Press, 2000.

———. *Legacies of the Turf: A Century of Great Thoroughbred Breeders*. Vol. 1. Lexington, KY: Eclipse Press, 2003.

———. *Matriarchs Volume II: More Great Mares of Modern Times*. Lexington, KY: Eclipse Press, 2008.

———. *Nashua*. Lexington, KY: Eclipse Press, 2001.

Bull, Phil. *The Best Horses of 1943*. London: Portway Press, 1944.

Canadian Thoroughbred

Canberra Times

Capps, Timothy T. *Spectacular Bid*. Lexington, KY: Eclipse Press, 2001.

Chronicle of the Horse

Clee, Nicholas. *Eclipse*. London: Transworld Publishers, 2009.

Daily Racing Form

Edwards, Elwyn Hartley. *Racehorse: The Complete Guide to the World of Horse Racing*. Hampshire, UK: Automobile Association Developments, 2008.

Gatto, Kimberly. *Belair Stud: The Cradle of Maryland Horse Racing*. Charleston, SC: The History Press, 2012.

Hewitt, Abram S. *Sire Lines*. Lexington, KY: Eclipse Press, 1977.

Hollingsworth, Kent. *The Kentucky Thoroughbred*. Lexington: University Press of Kentucky, 1976, 1985.

Horse Racing's Top 100 Moments. Lexington, KY: Eclipse Press, 2006.

Irish Horse. Vol. 18. Bloodstock Breeders'. Horse Owner's Association of Ireland, 1950.

Jodidio, Philip. *A Racing and Breeding Tradition: The Horses of the Aga Khan*. New York: Prestel USA, 2011.

Lennox, Muriel. *Dark Horse: Unravelling the Mystery of Nearctic*. Toronto, ON: Beech House Books, 2001.

Lexington Leader

Mitchell, Frank J. *Great Breeders and Their Methods: The Hancocks*. Neenah, WI: Russell Meerdink Company, Ltd., 2005.

Morris, Tony, and John Randall. *A Century of Champions*. Halifax, UK: Portway Press,1999.

———. *Horse Racing: Records, Facts, Champion*. 3rd ed. Enfield: Guinness Publishing, 1990.

Mortimer, Roger, Richard Onslow and Peter Willett. *Biographical Encyclopedia of British Flat Racing*. London: Macdonald and Jane's, 1978.

Nack, William. *Secretariat: The Making of a Champion*. Cambridge, MA: Da Capa Press, 1975, 2002.

Richards, Gordon. *My Story*. London: Hodder and Stoughton, 1955.

Sports Illustrated

Thoroughbred of California

Thoroughbred Record

Thoroughbred Times

Toby, Milton C. *Noor: A Champion Thoroughbred's Unlikely Journey from California to Kentucky*. Charleston, SC: The History Press, 2012.

INDEX

J

K

L

M

N

P

R

S

T

W

Z

ABOUT THE AUTHOR

Melanie Greene is a doctoral candidate in the Faculty of Education at Memorial University of Newfoundland. She has published numerous articles in peer-reviewed academic journals. She holds a BA and MA in sociology and has completed programs in equine studies at the University of Guelph. This is her first book.

Melanie has been an avid equestrian for much of her life. She has competed in dressage and hunter jumper divisions in her local horse community, as well as with the McMaster University Equestrian Team in Hamilton, Ontario.

Melanie currently resides in her hometown of St. John's, Newfoundland, Canada.